Alexander Reiterer

A Decision Support System for an Image Assisted Multi-Sensor System

Alexander Reiterer

A Decision Support System for an Image Assisted Multi-Sensor System

Concept and Implementation

Südwestdeutscher Verlag für Hochschulschriften

Impressum/Imprint (nur für Deutschland/ only for Germany)
Bibliografische Information der Deutschen Nationalbibliothek: Die Deutsche Nationalbibliothek
verzeichnet diese Publikation in der Deutschen Nationalbibliografie; detaillierte bibliografische
Daten sind im Internet über http://dnb.d-nb.de abrufbar.
Alle in diesem Buch genannten Marken und Produktnamen unterliegen warenzeichen-, marken-
oder patentrechtlichem Schutz bzw. sind Warenzeichen oder eingetragene Warenzeichen der
jeweiligen Inhaber. Die Wiedergabe von Marken, Produktnamen, Gebrauchsnamen,
Handelsnamen, Warenbezeichnungen u.s.w. in diesem Werk berechtigt auch ohne besondere
Kennzeichnung nicht zu der Annahme, dass solche Namen im Sinne der Warenzeichen- und
Markenschutzgesetzgebung als frei zu betrachten wären und daher von jedermann benutzt
werden dürften.

Verlag: Südwestdeutscher Verlag für Hochschulschriften Aktiengesellschaft & Co. KG
Dudweiler Landstr. 99, 66123 Saarbrücken, Deutschland
Telefon +49 681 37 20 271-1, Telefax +49 681 37 20 271-0, Email: info@svh-verlag.de
Zugl.: Wien, Technische Universität Wien, Dissertation, 2004

Herstellung in Deutschland:
Schaltungsdienst Lange o.H.G., Berlin
Books on Demand GmbH, Norderstedt
Reha GmbH, Saarbrücken
Amazon Distribution GmbH, Leipzig
ISBN: 978-3-8381-0352-5

Imprint (only for USA, GB)
Bibliographic information published by the Deutsche Nationalbibliothek: The Deutsche
Nationalbibliothek lists this publication in the Deutsche Nationalbibliografie; detailed
bibliographic data are available in the Internet at http://dnb.d-nb.de.
Any brand names and product names mentioned in this book are subject to trademark, brand or
patent protection and are trademarks or registered trademarks of their respective holders. The
use of brand names, product names, common names, trade names, product descriptions etc.
even without a particular marking in this works is in no way to be construed to mean that such
names may be regarded as unrestricted in respect of trademark and brand protection legislation
and could thus be used by anyone.

Publisher:
Südwestdeutscher Verlag für Hochschulschriften Aktiengesellschaft & Co. KG
Dudweiler Landstr. 99, 66123 Saarbrücken, Germany
Phone +49 681 37 20 271-1, Fax +49 681 37 20 271-0, Email: info@svh-verlag.de

Copyright © 2009 by the author and Südwestdeutscher Verlag für Hochschulschriften
Aktiengesellschaft & Co. KG and licensors
All rights reserved. Saarbrücken 2009

Printed in the U.S.A.
Printed in the U.K. by (see last page)
ISBN: 978-3-8381-0352-5

Contents

I Introduction   1

1 Introduction   3
   1.1 Motivation   3
   1.2 State of the Art   4
   1.3 Outline of this Work   6

II Basic Concept and Image Analysis   9

2 Concept of the Videometric Measurement System   11
   2.1 Introduction 11
   2.2 System Components 12
   2.3 Decision Process 14

3 Image Analysis   15
   3.1 Histogram Features 15
   3.2 Haralick Features 17
   3.3 Fuzzification / Abstraction of Image Features 18
   3.4 Other Image Features 20
   3.5 Summary 20

III Knowledge-Based System   21

4 Knowledge-Based Technology – Introduction   23
   4.1 Definition 23
   4.2 Knowledge-Based System Architecture 24
   4.3 Development Tools 28
   4.4 The Development Process 28
   4.5 Summary 29

5 Knowledge-Based Image Preprocessing and Image Enhancement System   31
   5.1 Basic Concept 31
   5.2 Image Preprocessing and Image Enhancement Algorithms 32
   5.3 The Knowledge Base – Image Preprocessing and Image Enhancement 40
   5.4 Examples 45
   5.5 Summary 48

6	**Knowledge-Based Point Detection**	**49**
	6.1 Basic Concept	49
	6.2 Interest Operators	50
	6.3 The Knowledge Base – Point Detection	55
	6.4 Examples	67
	6.5 Knowledge-Based Point Filtering	70
	6.6 Summary	76
7	**Image Matching Strategies**	**77**
	7.1 Introduction	77
	7.2 Raster-Based Matching	77
	7.3 Feature-Based Matching	80
	7.4 Relational Matching	81
	7.5 Summary	81

IV Implementation, Experiments and Alternative Technique 83

8	**Implementation**	**85**
	8.1 System Architecture	85
	8.2 The Prototype	90
9	**Experiments**	**93**
	9.1 First Example	93
	9.2 Second Example	96
	9.3 Summary	99
10	**Alternative Technique – Artificial Neural Networks**	**101**
	10.1 Introduction	101
	10.2 Experiments	103
	10.3 Summary	106

V Conclusion 109

11 Conclusion and Future Work — **111**

VI Appendix 117

A Haralick Features — **119**

B Membership Functions — **121**

C	**Knowledge Bases**	**123**
	C.1 Knowledge Base for Image Preprocessing and Image Enhancement	123
	C.2 Knowledge Base for Point Detection	133
	C.3 Knowledge Base for Point Filtering	139

Part I

Introduction

Chapter 1

Introduction

1.1 Motivation

In the past high precision online 3D measuring required artificial targets defining the points on the objects to be monitored. For many tasks like monitoring of displacements of buildings, artificial targets are not desired. At the *Institute of Geodesy and Geophysics* (*Vienna University of Technology*) a new kind of measurement robot based on modified videotheodolites has been developed within the last few years (Mischke, 1998; Roic, 1996). The fundamental idea of this system is to use the texture on the surface of the object to find *"interesting points"* by special image processing methods which can replace the artificial targets.

High precision on-line 3D measuring is required for many applications, among others:

- monitoring of displacements (buildings, produced workpieces, and others),
- quality control for production lines,
- hazardous site surveying.

The disadvantage of an on-line measurement system is the requirement for a well-trained "measurement expert" who has to have certain skills and experience to properly handle the complex system. From image capturing to deformation analysis a series of actions and decision making have to be performed. Reliable automatic or semi-automatic object surveying will be only possible if all the knowledge about the measurement system is available and included in a suitable program system. An approach to implement knowledge about a well defined problem domain is the development of knowledge-based systems, a particular class of computing applications.

In this work, we describe the development of a knowledge-based system which supports the operator when making the following decisions:

- selection of suitable image preprocessing and image enhancement algorithms,
- selection of suitable point detection algorithms (*interest operators*) and
- point filtering.

Due to the complexity of these tasks, fully automatic decision-making is not operational. For this reason our approach is an automatic decision-system with integrated user-interaction.

As mentioned above the point finding process bases on the texture of the surface. Therefore well textured and structured objects are qualified for such a measurement system. There exist a lot of objects which fulfil this demand. To restrict the development process our measurement system is focused on monitoring of building facades. Despite offering a wide range of structures most facades can be represented by simple line geometry. Integrated knowledge, examples and simulations are fitted according to this. An extension to other objects, like car bodies or aircraft is possible and envisaged.

In Section 1.2, an overview of the state of the art in the field of *videotheodolite measurement systems* and in the field of *knowledge-based systems* will be given.

1.2 State of the Art

1.2.1 Videotheodolite Measurement Systems

Measurement robots have been used in various surveying applications for several years. The first one, the *GeoRobot*, was a modified theodolite constructed by Kahmen and used for monitoring ground movements of opencast mine slopes (Kahmen et al., 1983). Videotheodolites, the combination of CCD camera and motorized theodolite, have been working successfully in several areas for nearly twenty years (most notably: aircraft and ship building industry). The disadvantage of this system was the dependence on active targets or retroreflectors. To overcome these restrictions a new research project was started in 1990 (*"Optical 3D sensing of static and moving scenes by computer-controlled rotating cameras and 3D information extraction"* - research program *"Theory and applications of digital image processing and pattern recognition"*), with the main goal of developing a robot measurement system for object recognition, using not only artificial targets but also the surface texture by scanning the object automatically with videotheodolites. The developed measurement system consists of different components: videotheodolites, components for image acquisition and image processing, a component for system control, and some output devices.

Fabiankowitsch (1990) was the first researcher who experimented with the new measurement system. His work resulted in special measurement methods (filter techniques for active targets).

In 1994, the research project *"Stereovideometry and Spatial Object Recognition"* commenced. One of the milestones of this main research was the work done by Roic (1996). The aim of his work was to prepare images of unsignalized targets by using image processing methods, in order to make interactive and automatic spatial surface measurement possible.

A few years later Mischke (1998) developed a powerful measurement system based on two videotheodolites. The system was able to measure active or passive targets and non-signalized points, like intersections of edges or lines. Mischke has implemented the "Förstner interest operator" (Förstner, 1991) to select all remarkable but non-signalized points. *Interest operators* were well-known for offline applications in photogrammetry, but had not been used for videotheodolites before.

In 2001 the research project *"Theodolite-based and Knowledge-based Multi-Sensor-System"* began, with the main goal of developing a knowledge-based measurement system for 3D-sensing of static or moving scenes by computer-controlled rotating cameras and 3D information extraction. The work at hand was created in the course of this project.

1.2 State of the Art

De Seixas (2001) has developed a scanning method based on different grid-line methods. The measurement system was expanded by a laser-theodolite which is coupled to a laser generator of a visible laser beam that projects target points onto the object. The videotheodolite is automatically pointed at the same time as the pointer-theodolite. The videotheodolite identifies the target point and determines its position on the CCD array (Kahmen et al., 2001). The work researched by De Seixas was continued and refined by Von Webern (2003).

Hovenbitzer (2001) has described the realization and the calibration of two different measurement systems, which are designed for measuring contactless 3D data in close range. He compared the *Hybmess* system, which is based on a standard motorized tacheometer extended by a CCD-line camera, and the *3dLMS* system, which determines the co-ordinates of surface points by reflectorless distance measurement. Additionally he presented a method for processing and visualization of measured data.

Niessner (2002) has developed a colour-based segmentation method which can be used as a qualitative measuring module. Different colours represent the starting point for the distinction of objects in the image. The method developed extracts individual objects and analyses their deformation. The image information is used for a qualitative deformation analysis.

During last year, research in the area of videotheodolites increased. *Leica Geosystems* developed in 2003 a prototype of an image-assisted total station. The purpose of the work done by *Leica Geosystems* is to define a hybrid or semi-automatic way to combine the strength of the traditional user-driven surveying mode with the benefits of modern data processing (Walser et al., 2003). *Leica Geosystems* has modified a Total Station TCRA 1101 to a prototype of an image-assisted total station.

Furthermore, at the *Intergeo 2002*, Sokkia introduced the tacheometer SET3110MV study (prototype), which enables focused color images.

At the *Technische Universität München* Wasmeier (2003) has developed a measurement system for object recognition based on the Leica tacheometer TCA2003. The system was developed to automatically recognize steeples and to point the collimation axis of the instrument towards the target point.

1.2.2 Knowledge-Based System

During the first decade of *Artificial Intelligence*, research problem solving was of a general-purpose search mechanism trying to string together elementary reasoning steps to find complete solutions. These methods use only weak information about the problem domain. The way around this weak performance is to use knowledge more suited to making larger reasoning steps and to solving typically-occurring cases in narrow areas of expertise (Russell et al., 1995).

An early example of this approach is DENTRAL. It was developed at Stanford university by Ed Feigenbaum, Bruce Buchanan and Joshua Lederberg to solve the problem of inferring molecular structure from the information provided by a mass spectrometer (Russell et al., 1995). The next major effort was the development of MYCIN to diagnose blood infections. MYCIN, having about 450 rules, was able to perform as well as some experts. A series of approaches to medical diagnostic and to other academic disciplines followed (PROSPECTOR, LUNAR, R1, and others).

In recent years, various knowledge-based systems were developed and implemented in geodesy. The following works are worth mentioning:

In 1995 the University of Hannover carried out the research project SAMBA (System zur Anwendung der Messtechnik im Bauwesen) to develop a knowledge-based system for measurement engineering especially for bridges (Kuhlmann, 1993; SAMBA-Endbericht, 1995). A full expert-system shell was developed and quite a number of rules were implemented.

Brezing (2000) developed an expert system for deformation analysis, including analysis and interpretation of deformations. A special aspect of this work is dedicated to the modelling of the knowledge. He implemented the knowledge base as rule and object based approach. For this development no expert system shell was used; the system was built (in Lisp) from scratch.

Reiterer (2001) developed a prototype of a motorized digital level including a control software enlarged by a knowledge-based system, which optimises the monitoring by means of complete information. Its suitability in practice was tested by detailed test measurements and one long time monitoring.

Chmelina (2002) presented the concept and prototype of a knowledge-based software for automatic detection of significant 3D displacement behaviour in tunneling. By then the interpretation of the displacements was mainly based on the manually-executed inspection of numerous and different types of displacement diagrams.

A large number of other applications of knowledge-based systems in geodesy and geotechnique are listed in (Chmelina, 2002).

1.3 Outline of this Work

This work has to be seen in the context of a videotheololite based 3D measurement system as it will be described in Chapter 2.

Part II is dedicated to the basic concept of the developed measurement system and image analysis. In this work, image analysis represents the basis for all the automatic or semi-automatic decision-making processes. To understand the implemented decision system, it is necessary to know some details about image analysis algorithms (Chapter 3).

Part III deals with the development of the knowledge-based system. In this part we will describe the three knowledge-based systems which help the user in several decisions:

- a knowledge-based system for choosing an algorithm, a group of algorithms, or a combination of algorithms for image preprocessing and image enhancement;
- a knowledge-based system for choosing a suitable *interest operator* for finding measurement-points;
- a knowledge-based system for point filtering.

In part IV, we will present the developed prototype, which includes all techniques and methods developed in the course of the work at hand. Additionally, in this part we will present the results of some experiments done off-line. An off-line test suite is more flexible than an on-line one. Tests can be done quicker, and a greater number of test cases is possible. Part IV will close with a description of an alternative technique to knowledge-based systems: *neural networks*. Having described the theoretical foundations, the realized networks will be explained by means of experiments.

1.3 Outline of this Work 7

Finally, part V will present some conclusions of this work, and will give an outlook on future work to be done to increase the degree of automation of the measurement system and to improve the whole system.

Part II

Basic Concept and Image Analysis

Chapter 2

Concept of the Videometric Measurement System

2.1 Introduction

A central topic of videometrics is the calculation of three-dimensional object-point coordinates from two-dimensional image (pixel) coordinates. Different sensors and techniques can be used depending on the object to be reconstructed.

Applications of 3D reconstruction can be: mapping, architecture, construction, medicine, industrial applications, and others.

A videometric measurement system (Figure 2.1) consists of a light source, image sensor, components for image acquisition and image processing, a computer for system control and some output devices.

Kahmen et al. (2001) give a definition for *videometry*:

> The use of devices for optical sensing to automatically receive and interpret an image of a real scene, to obtain geometrical information, and to control direction information, the particular location, or orientation of a part in an assembly, the simple presence or absence of an object, or a part of an assembly.

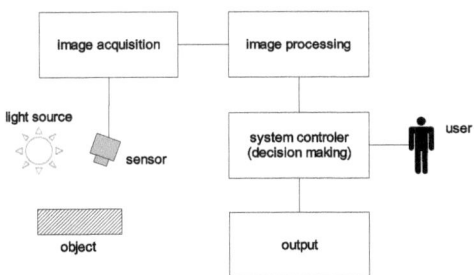

Figure 2.1: Concept of a videometric imaging system (Kahmen et al., 2001).

The concept of our measurement system is based on videometry and was developed by Roic (1996), Mischke (1998) and Kahmen et al. (2001) in a two-step development process:

First step: The key element of the first system was image processing software that supports the operator to find "natural targets". The operator has to choose the image processing steps and to analyse whether the processed images can be used as targets. The result of this first development step was a non-automatic interactive measurement system (Roic, 1996; Kahmen et al., 2001).

Second step: The objective of the second step was to develop an automatic measurement system. This was realised by using the videotheodolites in a master and slave mode. The master-theodolite scans the object while the slave-theodolite tracks it by automatically searching for homologous regions. Two scanning methods were developed: scanning with the *Förstner-interest operator* and scanning with different *grid-line methods*. The result of this second development step was the semi-automatic measurement system shown in Figure 2.2 (Mischke, 1998; Kahmen et al., 2001; De Seixas, 2001).

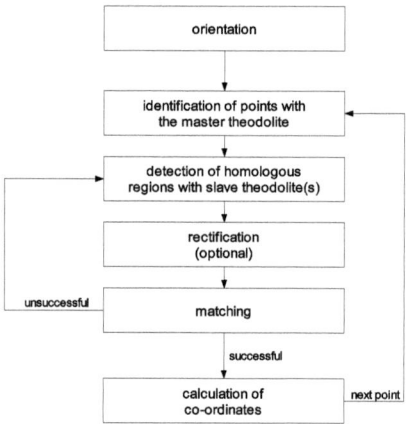

Figure 2.2: Diagram of the semi-automatic measurement system developed by Mischke (1998).

The main goal of the current development step (the *third development-step*) is to automate different decision makings by the measurement system. This is done by the integration of a decision system. The complete measurement system will be described in the next section.

2.2 System Components

The complete measurement system (Figure 2.3) is a combination of different components:
- sensors (videotheodolites used as image sensors),
- a computer system,
- software (control system, image processing and decision system),
- accessories.

2.2 System Components

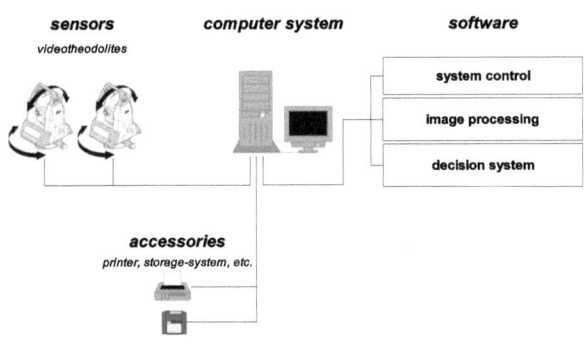

Figure 2.3: Complete measurement system.

Image sensors used to capture data are two videotheodolites Leica TM3000V/VD. A videotheodolite has a CCD camera in its optical path. Images of the telescope field are projected onto the camera's CCD chip. It is possible to project the images from the telescope's image plane to the CCD array or to switch to a special wide-angle optical system ($9 \times 12°$) to produce a general view of the object. The wide-angle view is only used for approximate target finding. The horizontal and vertical axes carrying the telescope and the CCD cameras are driven by motors, which are controlled by a computer. More details about the image sensors used can be found in (Fabiankowitsch, 1990; Roic, 1996; Mischke, 1998). It should be noted that for all tests described in this work, we have additionally used two conventional digital cameras (Canon G2 and Canon MV6i). The resolutions of the captured images are 640×480 and 2272×1704 pixels. The use of conventional digital cameras is more flexible than the use of videotheodolites, since they are easier to carry than the whole measurement system needed for the use of videotheodolites. For this reason with conventional digital cameras image capturing can be done quicker and in a greater number.

The *computer system* used is a conventional personal computer (AMD Athlon – 1.5 Ghz – 256 MB Ram) with integrated PCI frame grabber (Data Translation DT3131). A frame grabber is needed to capture a single frame (image) from the analog video signal (from image sensors) and stores it as a digital image under computer control.

The developed *software* is divided into three parts: software for system control, software for image processing (image preprocessing and image enhancement, *interest operators*, image analysis, and others) and the decision system.

The main steps of the developed on-line measurement system are:

- image capturing,
- image preprocessing and image enhancement,
- detection of *interest points* by different *interest operators*,
- detection of homologous points with the slave-videotheodolite,
- calculation of co-ordinates,
- point filtering,

- point-matching (between different epochs),
- deformation analysis.

2.3 Decision Process

As we have explained, the main goal of the current development step is the integration of a suitable decision system. A decision systems is a specific class of a computerized information system that supports engineering, business or organizational decision-making activities. A properly designed decision systems is an interactive software-based system intended to help decision makers to solve problems and to make decisions. Such a system consists in its plainest form of three main components, namely an input component, decision algorithm(s) and an output component (Figure 2.4).

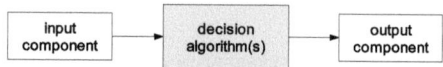

Figure 2.4: Decision system.

In our system all decisions have to be done on the basis of the captured image or on values which represent this image. Using the whole image as input is, because of processing time, not suited for an on-line measurement system. For this reason we use appropriate values as input for the decision system. The process for extracting these values from the image is called *image analysis* and will be content of the next chapter.

Chapter 3

Image Analysis

In this work the goal of image analysis (image feature extraction) is to extract information needed as input for the decision system. For our application, image analysis is one of the most critical tasks and bottlenecks in the processing chain.

There exist many approaches for image analysis, which can roughly be divided into three categories: structural, spectral and statistical methods (Pratt, 1978).

Structural techniques characterise texture being composed of simple primitives called "texels" (texture elements), that are regularly arranged on a surface. *Spectral techniques* are based on properties of the Fourier spectrum and describe global periodicity of the grey-levels of a surface by identifying high energy peaks in the spectrum. *Statistical techniques* characterise texture by the statistical properties of the grey-levels of the points comprising a surface. These properties are computed from the grey-level histogram by simple mathematical routines. *Statistical techniques* have low calculation effort and are therefore suitable methods for an on-line system, by which fast execution of image analysis is necessary. To satisfy this demand and to represent the image properties in a suitable form for the subsequent decision process we use the following *statistical techniques*:

- histogram feature extraction and
- Haralick feature extraction.

3.1 Histogram Features

The histogram of an image is a plot of the grey-level values versus the number of pixels at that value. It can be utilized to generate a class of image features (histogram features). The first-order probability distribution of image amplitude may be estimated from:

$$h(z) = \frac{H(z)}{N}; \qquad \sum_{z=z_u}^{z_o} h(z) = 1. \qquad (3.1)$$

N represents the total number of pixels in the full image and $H(z)$ the number of pixels of amplitude z. Based on the histogram, several features have been formulated. *Mean (M_1), standard deviation (M_2) and skewness (M_3)* describe the shape of the image histogram.

The shape of an image histogram provides a large amount of information about the character of the image. A narrowly distributed histogram indicates a low-contrast image while a bimodal histogram suggests regions of different brightness (Pratt, 1978). An example is given in Figure 3.1.

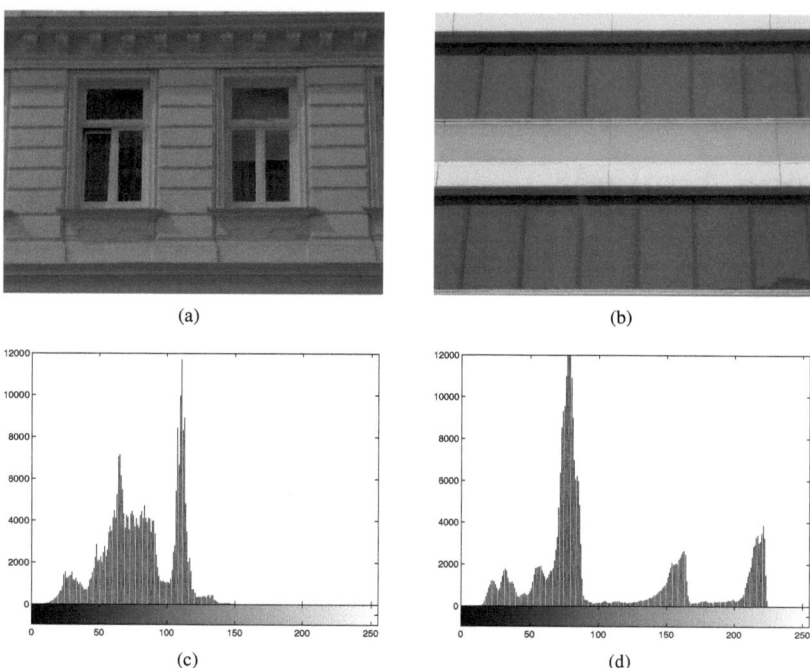

Figure 3.1: (a) A low-contrast image and (b) an image with three regions of different brightness, including their histograms (c, d).

The first histogram feature calculated is *mean* (M_1) by dividing the sum of a set of grey-levels by the number of grey-levels. In our case the set enfolds the values of the full image. *Mean* is correlated to the brightness of the image: a bright image will have a high mean and a dark image will have a low mean.

Mean is defined as:

$$M_1 = \overline{z} = \sum_{z=z_u}^{z_o} z h(z). \tag{3.2}$$

Variance (M_2) is a measure of the average distance between each of a set of grey-levels and their mean value. The *standard deviation* is the square root of the *variance*. It describes the spread of the grey-levels; a high-contrast image will have a high *standard deviation* and a low-contrast image will have a low *standard deviation*.

Variance is defined as:

$$M_2 = \sum_{z=z_u}^{z_o} (z - \overline{z})^2 h(z). \tag{3.3}$$

Skewness (M_3) is a measure of the symmetry of distribution of grey-levels around its mean. A distribution of grey-levels is symmetrical if the histogram looks the same to the left and right of the center point; any symmetrical data should have a skewness near zero. Negative values for the skewness indicate data that are skewed left and positive values for the skewness indicate data that are skewed right. *Skewness* tells about a balance of the bright and dark areas in the image. The *skewness* value will be positive for an image with stronger dark areas and negative for an image with stronger bright areas.

Skewness is defined as:

$$M_3 = \sum_{z=z_u}^{z_o} (z - \overline{z})^3 h(z). \tag{3.4}$$

3.2 Haralick Features

Haralick et al. (1993) proposed 13 measures of textural features which are derived from the co-occurrence matrices, a well-known statistical technique for texture feature extraction. Texture is one of the most important defining characteristics of an image. the grey-level co-occurrence matrix is the two dimensional matrix of joint probabilities $p(i,j)$ between pairs of pixels, separated by a distance d in a given direction r. It is popular in texture description and builds on the repeated occurrence of some grey-level configuration in the texture. Figure 3.2 shows the generation of co-occurrence matrices for an image with four possible pixel intensity values (0 – 3). Generation will start in the top left corner and count the occurrences of each reference pixel to neighbouring pixel relationships. In our example, it can be seen that for $r = 0°$ there were four occurrences of 2 to the left of 1.

In this work we use only some of the 13 Haralick features: the *angular second moment* (H_1), the *contrast* (H_2), the *inverse difference moment* (H_5) and the *entropy* (H_9). A listing of all the 13 Haralick features can be found in Appendix A, a detailed description in (Haralick et al., 1993).

The *angular second moment* (H_1) is a measure of the homogeneity of an image. Hence it is a suitable measure for the detection of disorders in textures. For homogeneous textures, the value of angular second moment turns out to be small compared to non-homogeneous ones.

The *angular second moment* is defined as:

$$H_1 = \sum_i \sum_j p(i,j)^2. \tag{3.5}$$

The *contrast* (H_2) is a measure of the amount of local variations present in an image. This information is specified by the matrix of relative frequencies $p(i,j)$ with which two neighboring pixels occur on the image, one with grey value i and the other with grey value j.

The *contrast* is defined as:

$$H_2 = \sum_{n=0}^{N_g-1} n^2 \left\{ \sum_{i=1}^{N_g} \sum_{j=1}^{N_g} p(i,j) \right\}; \quad n = |i - j|. \tag{3.6}$$

The *inverse difference moment* (H_5) measures image homogeneity too. It achieves its largest value when most of the occurrences in co-occurrence matrices are concentrated near the main diagonal. The *inverse difference moment* is inversely proportional to the contrast of the image.

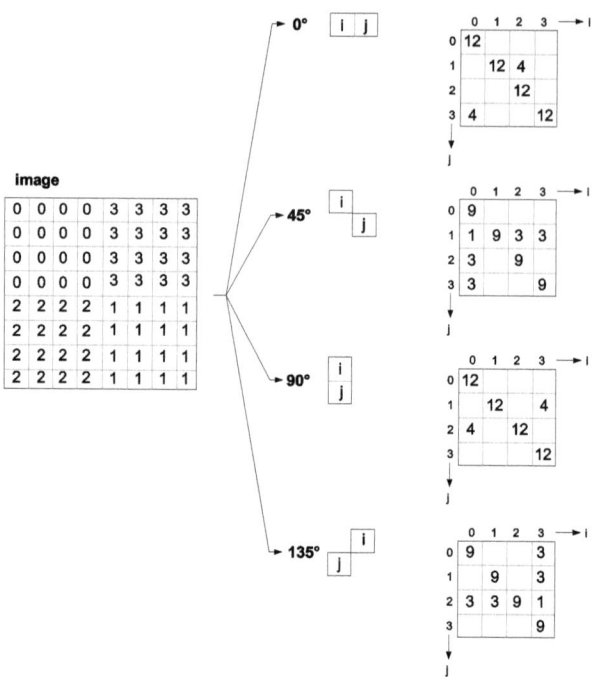

Figure 3.2: Generation of co-occurrence matrices for a distance $d = 1$ and given directions $r = 0°$, $45°$, $90°$, $135°$ (Bässmann et al., 1998).

The *inverse difference moment* is defined as:

$$H_5 = \sum_i \sum_j \frac{1}{1 + (i-j)^2} p(i,j). \tag{3.7}$$

The *entropy* (H_9) gives a measure of complexity of the image and is related to the information-carrying capacity of the image. *Entropy* is maximized when the probability of each entry is the same, thus a high value for entropy means that the grey-level changes between pixels are evenly distributed, and the image has a high degree of visual texture.

The *entropy* is defined as:

$$H_9 = \sum_i \sum_j p(i,j) \, log \, \{p(i,j)\}. \tag{3.8}$$

3.3 Fuzzification / Abstraction of Image Features

To make the extracted image features more suitable for the knowledge-based decision system we use a special fuzzification / abstraction procedure.

3.3 Fuzzification / Abstraction of Image Features

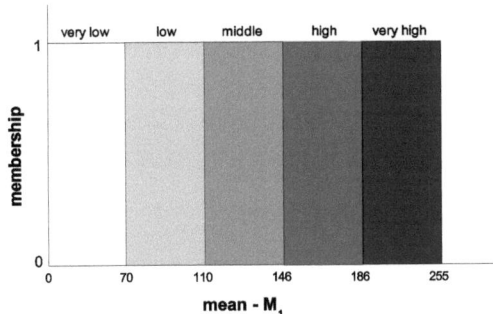

Figure 3.3: Fuzzification / abstraction of the histogram feature *mean*.

This procedure translates the input values (image features) into linguistic concepts, which are represented by abstraction ("fuzzy") sets. Figure 3.3 considers an abstraction set for the histogram feature *mean* with five members: very low, low, middle, high and very high.

This technique is not a fuzzification in terms of definition. We use only non-overlapping spring membership functions. The use of such a abstraction procedure permits us to write rules in terms of easily-understood word descriptors, rather than in terms of numerical values.

All membership functions for our fuzzification / abstraction are listed in Appendix B.

Extracted image features and "fuzzy-values" (f.v.) of the two images of Figure 3.1a and 3.1b are shown in Table 3.1 and Table 3.2.

M_1	M_1 - f.v.	M_2	M_2 - f.v.	M_3	M_3 - f.v.		
79.7320	low	26.2376	low	-0.1946	very low negative		
	0°	f.v.	45°	f.v.	90°	f.v.	135°	f.v.	average	f.v.
H_1	0.003	high	0.002	mid.	0.002	mid.	0.002	mid.	0.002	mid.
H_2	32.280	v.low	65.775	low	37.696	v.low	67.715	low	50.866	low
H_5	0.504	high	0.347	mid.	0.415	high	0.337	mid.	0.401	high
H_9	2.941	low	3.216	mid.	3.087	mid.	3.229	mid.	3.118	mid.

Table 3.1: Extracted image features for Figure 3.1a.

M_1	M_1 - f.v.	M_2	M_2 - f.v.	M_3	M_3 - f.v.		
104.9926	low	56.9849	middle	0.8757	very high positive		
	0°	f.v.	45°	f.v.	90°	f.v.	135°	f.v.	average	f.v.
H_1	0.0039	high	0.0031	high	0.0038	high	0.0032	high	0.0035	high
H_2	12.569	v.low	157.067	mid.	148.027	mid.	160.944	mid.	119.652	mid.
H_5	0.5780	high	0.4554	high	0.5107	high	0.4646	high	0.5022	high
H_9	2.7539	low	3.0044	middle	2.9175	low	2.9939	low	2.9174	low

Table 3.2: Extracted image features for Figure 3.1b.

3.4 Other Image Features

As mentioned above, the application of the developed measurement system is focused on building displacement-monitoring, especially on the monitoring of facades. In addition to the extracted image features (histogram features and Haralick features), further information (of "non measurable" nature) about the object is collected by different user-queries.

The user has to answer the following questions (UQ):

- UQ1: What kind of type is the facade?
- UQ2: Are there any reflections on the facade?
- UQ3: Is there a shadow cast on the facade?

For the first question, four answers are available: old building facade (type A), new building facade (type B), brick-lined facade (type C) and steel-glass facade (type D). We have created these four facade types since most of existing buildings (for the central european culture) can be characterized by these types. For the other two questions the user has the possibility to characterise the strength of existing effects (none, slight, middle, strong, very strong).

Haralick features (*angular second moment*, *inverse difference moment* and *entropy*) are calculated from each co-occurrence matrix ($d = 1$ and $r = 0°, 45°, 90°, 135°$) and their values (including average values) are saved together with the histogram features (*mean*, *variance*, *skewness*) and the user-queries in a special data-file. In addition to the crisp values, fuzzy values of all features are stored. The whole file is used as an input-file for the knowledge-based decision system. More about the file structure can be found in Section 8.1.1.

3.5 Summary

In this chapter we have presented a method for image analysis. It bases on statistical properties of the grey-levels and consist of two different techniques, namely histogram feature extraction and Haralick feature extraction. Additionally to these image features we collect "non measurable" image information by user questions.

All these values will be used as input for the decision system (Figure 2.4). The next step has to be the development of the decision algorithms. In informatics there exist various approaches for algorithm and software development (module-oriented, object-oriented, knowledge-based, neural networks, software agents, and others). The conventional approach to software development is challenged by new ideas. In the following we will use a knowledge-based approach. This technique has been developed more than thirty years ago and has some exceptional advantages in comparison with conventional approaches. More about the advantages and an introduction to this "new" technique, understandable even for non computer scientists, in the next chapter.

Part III

Knowledge-Based System

Chapter 4

Knowledge-Based Technology – Introduction

As mentioned above, we have chosen a knowledge-based approach for the decision system. Knowledge-based technology has been an active area of research for more than thirty years (Russell et al., 1995), and is currently absorbing a significant percentage of product development capacity, not only in universities but also in corporations and government. It has attracted particular attention within competitive, information-dependent industry sectors, such as financial institutions, business consultancies or information centres.

In this chapter we will give an overview of knowledge-based technology and we will emphasize the advantages of this technique in comparison with a conventional approach.

4.1 Definition

Gaining a clear understanding of the technology can be difficult, because key terminology is used in ways that are subtly different, sometimes significantly so. The term "knowledge-based system" has been defined along the following lines:

> *A computerized system that uses knowledge about some domain to arrive at a solution to a problem from that domain. This solution is essentially the same as that concluded by a person knowledgeable about the domain of the problem when confronted with the same problem* (Gonzalez et al., 1993).

The main characteristics of knowledge-based systems are:

- simulating human reasoning about a problem domain,
- reason using representations of knowledge,
- solving problems using heuristics or rules of thumb,
- able to explain and justify solutions in order to convince user that reasoning is valid.

Following strictly the definition and the characteristics above, various conventional software could be incorrectly categorized as knowledge-based systems, e.g. programs that analyze the stress factors of buildings or analyze geotechnical measurements. These programs perform the same analysis as an expert in the field of knowledge, but they are not knowledge-based systems.

Gonzalez et al. (1993) has formulated three fundamental concepts which distinguish knowledge-based systems from conventional algorithmic programs and from general search-based programs:

- the separation of the knowledge from how it is used,
- the use of highly specific knowledge,
- the heuristic rather than algorithmic nature of the knowledge employed.

The first underlying idea is the most important and leads to one of the biggest advantages of knowledge-based systems compared to conventional software. The fundamental concept of the separation of domain-knowledge from reasoning mechanism makes it easy to modify or extend the knowledge. In modern information technology changes occur frequently and knowledge has to be updated; ease of modification is a very important feature in these situations.

4.2 Knowledge-Based System Architecture

Knowledge-based systems are complex computer software products that can be viewed from the perspective of an end user or from the perspective of a knowledge engineer. From the point of view of the user a knowledge-based system consists of an intelligent program, a user interface and problem-specific database. The system is a black box that operates based on unknown algorithms. From the developer's (knowledge engineer) view the system consists of two components: an intelligent program and the development shell. The development shell is a set of tools that assist the implementation of relevant knowledge.

Independently from the point of view, knowledge-based systems consist of the following major components:

- a knowledge base,
- an inference engine,
- an user interface,
- a knowledge acquisition tool,
- an explanation tool.

As we have explained above one fundamental characteristic of knowledge-based systems is the clear and clean separation between the domain knowledge and the problem solving knowledge. The two components that compose the "intelligent program" are the knowledge base and the inference engine. Along with the user interface they form the core of the system. The coaction of the different components is shown in Figure 4.1.

4.2.1 Knowledge Base

The knowledge base, the most important component of a knowledge-based system, contains the relevant domain-knowledge that the knowledge engineer has implemented in the course of the development process (see Section 4.4).

4.2 Knowledge-Based System Architecture

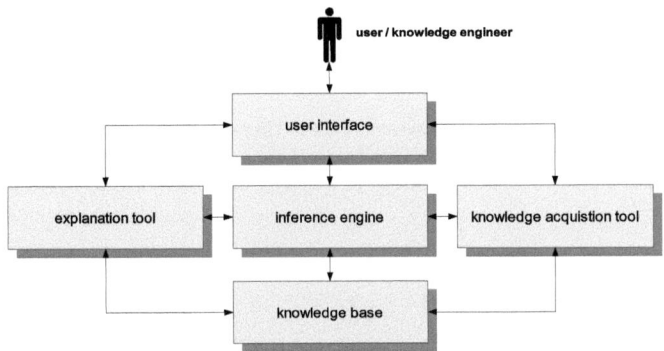

Figure 4.1: Architecture of a knowledge-based system (Gottlob et al., 1990).

We differentiate between *procedural* and *declarative* knowledge. *Procedural* knowledge is typically incorporated in an algorithm or a program, whereas *declarative* knowledge is represented explicitly and symbolically and should be independent of the methods to perform inferences on it.

Several declarative knowledge representation schemas are commonly used: predicate logic, production rules, semantic nets, frames and others. Each representation form has different advantages and disadvantages and is more or less suitably for a specific problem.

The knowledge engineer has to decide in favor of a suitable representation form in the course of knowledge implementation.

Our knowledge base was realised as *rule-based / object-oriented* approach. *Rule-based* programming is one of the most commonly used techniques for implementation of a knowledge base. Rules are used to represent heuristics, or "rules of thumb", which specify a set of relationships.

The structure of a *rule-based* approach is very similar to the way how people solve problems. Human experts find it convenient to express their knowledge in the form of rules (situation–action pairs). Further rules are a way to represent knowledge without complex programming constructs.

A *rule-based* approach consists of two parts:

- a set of rules,
- a working memory.

A **rule** is composed as an *if–then*–statement in conventional programming. The following examples are given for better understanding in pseudo code. The syntax of concrete implementation will be described in Chapter 8. We consider the following example rule in order to explain relevant concepts:

```
example_rule
   precondition one
   precondition two
   ...
⇒
   actions
```

A rule is divided into two parts, namely the *lefthand side* (LHS) and the *righthand side* (RHS) in our example with "⇒" separating both parts. In the LHS, we formulate the preconditions of the rule, whereas in the RHS, the actions are formulated. A rule can be applied (or *fired*) if all its preconditions are satisfied; the actions specified in the RHS are then executed.

The second component of a *rule-based system*, the **working memory**, is a collection of *working memory elements*, which itself are instantiations of a *working memory type* (WMT). WMTs can be considered as `record` declarations in PASCAL or `struct` declarations in C. An example of a WMT is as follows:

```
working_memory_type_example
  nr  (type INTEGER)
  M1  (type FLOAT)
  M2  (type FLOAT)
```

A *working memory type* is a user-defined composite type. It is composed of fields or members that can have different types. Our example has the name "working_memory_type_example" and consists of three fields (in program languages for knowledge-based system these fields are called often *slots*): nr, M1, M2. Each field is defined through a name and a type. Possible types are: integer, float, symbol and others.

4.2.2 Inference Engine

The phase where all rules are checked against all working memory elements is called *matching-phase*. The process of matching is accomplished by the inference engine. The result of the matching phase is the *conflict set*, which includes all *rule instances* "ready to be fired". A *conflict resolution strategy* selects one *rule instance* which is actually fired.

The process of deriving a solution to a problem can be viewed simplistically as one of finding a connection between the input and a conclusion. The creation of these connections can be done in several ways:

- from the features to the solution (*forward reasoning*),
- from the solution to the features (*backward reasoning*).

Although there are various forms for representing knowledge, it should be obvious that the inference engine must support the representation scheme used by the knowledge engineer. Details of these reasoning schemes are provided by Gonzalez et al. (1993). In the following, we will describe the *forward reasoning inference process*, the primary inference method of our knowledge representation language.

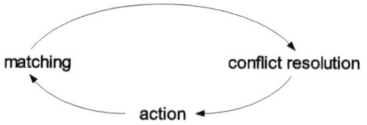

Figure 4.2: Recognize–act–cycle.

4.2 Knowledge-Based System Architecture

A *forward reasoning* inference engine works in a so-called "recognize–act–cycle" (Figure 4.2):

- it recognizes (by matching) which rules in the knowledge base can be satisfied by the information in the *working memory*,
- it decides which of the applicable rules should be used ("conflict resolution"),
- it applies the rule (using inferencing) and adds information to the *working memory* or deletes an earlier item from *working memory*, then goes back to the first step.

The process stops if all of the applicable rules have been executed, (the *conflict set* is empty) or the answer was found.

4.2.3 User Interface

The user interface (UI) serves to provide the end user and the knowledge engineer with a friendly means of communication with the program system. The user interface of a knowledge-based system can be used for different purposes like displaying the derived results or allowing the user to query the knowledge-based system as to why or how a particular decision was made (Gonzalez et al., 1993). There are several types of user interfaces: graphical user interface (GUI), text-based user interface, user interface based on natural language, and others. The most common type of user interface is the graphical user interface.

In the development phase of the user interface the end users have to be consulted about their needs and expectations. Knowledge-based systems and conventional program systems have been known to fail due to inadequate design of their user interface.

For our knowledge-based system we have implemented a graphical user interface. Chapter 8 describes the functionality and gives some screenshots.

4.2.4 Knowledge Acquisition Tool

The knowledge engineer has to interact with the domain specialist to acquire knowledge about the problem domain and to implement this knowledge in the knowledge base. The knowledge acquisition tool assists the knowledge engineer in this work. A simple form of a knowledge acquisition tool is a conventional text editor, like Wordpad or Emacs, which provides a view of the implemented knowledge and allows the knowledge engineer to edit the knowledge base. In a sophisticated form, this tool provides a wide range of features: debugging, bookkeeping functions, search functions and others. A well-engineered knowledge acquisition tool can significantly shorten the development phase.

4.2.5 Explanation Facility

The explanation facility helps the user to understand the knowledge reasoning of the system. Explanations (automatically or if required) enhance the usefulness and acceptance of knowledge-based systems.

An explanation facility normally allows the user to ask the system *why* a special solution was found and *how* the system reached its conclusions. An alternative solution is to document the *why* and *how* of every conclusion in a suitable file.

4.3 Development Tools

Tools and techniques for the development of knowledge-based systems are quite different from those of conventional program development. Generally, there are two methods for developing a knowledge-based system (Gonzalez et al., 1993): using a "tool", or developing from scratch. For a well-defined problem, the availability of a suitable development tool should be investigated before deciding on the development of the knowledge-based system from scratch. Using a pre-existing tool saves time and effort.

A development tool is a software package that contains the main components and features needed for developing a knowledge-based system. We refer to these as *knowledge-based shells* (simply *"shells"*). Numerous *shells* are commercially and non-commercially available. Worth mentioning are the CLIPS system developed by the NASA and the ILOG-RULES system developed by the ILOG company.

ILOG-RULES system is a complete development environment, including rule editor, rule management tool and a integrated graphical development tool for developing and debugging rules. More details about ILOG-RULES can be found in ILOG-White Paper (2003).

CLIPS is a productive development tool which provides a complete environment for the construction of rule and object based knowledge-based systems. The origins of the C Language Integrated Production System (CLIPS) date back to 1984 at NASA's Johnson Space Center. At that time nearly all knowledge-based system tools used LISP, a list-processing language, as the base language. Several problems hindered the use of LISP based development tools within NASA: the low availability of LISP development environment on conventional computers, the high cost of LISP tools, and the poor integration of LISP in and with other program languages (especially C and C++). Because of its portability, extensibility, capabilities, and low cost, CLIPS has received widespread acceptance throughout the government, industry, and academia. CLIPS is now maintained independently from NASA as public domain software CLIPS (2004).

Our knowledge-based system was realised in CLIPS. More details about the implementation can be found in Chapter 8.

4.4 The Development Process

During the development phase (Figure 4.3), knowledge is extracted from one or more persons who have specialised knowledge in the relevant domain. Such a person is usually called "expert" or "domain specialist". The knowledge is commonly expressed in the form of *"if-then-statements"*.

In some cases it may be possible for the domain specialist to feed the knowledge directly into the knowledge base, but more usually a knowledge engineer (analyst/programmer) captures it using special tools and techniques. This procedure is called *"knowledge engineering"*.

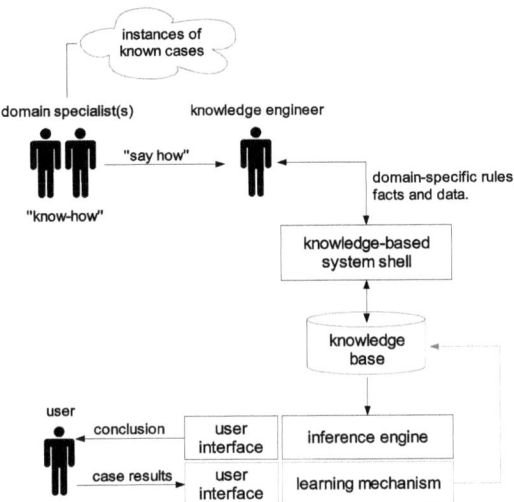

Figure 4.3: Model of development of a knowledge-based system.

All too often the *knowledge engineering* is seen as a transfer process of human knowledge into an implemented knowledge base. This transfer is based on the assumption that the knowledge which is required by the knowledge-based system already exists and just has to be collected and implemented. The required knowledge was obtained by interviewing experts on how they solve specific tasks.
Nowadays the process of building a knowledge-based system may be seen as a modeling activity. Building a knowledge-based system means building a computer model with the aim of realizing problem-solving capabilities comparable to a domain expert.
In our case the domain specialist has implemented the domain knowledge directly into the knowledge base using an appropriate language and supporting software.

In the use-phase the user, without reference to the domain specialist, consults the knowledge base. He provides information about some event or situation within the problem domain. The knowledge-based system (the inference engine) applies the rules stored in the knowledge base to the case-specific data. A result is provided to the user, in the form of a diagnosis, prognosis or decision, depending on the nature of the application. In addition the results of new cases can be used by the software (learning mechanism) to modify or to expand the existing knowledge base.

4.5 Summary

In this chapter we have explained the basics of knowledge-based techniques. In comparison with conventional software, the separation of the knowledge from how it is used, leads to the most important advantages of knowledge-based systems: easy modifiable and extensible. In modern information technology knowledge has to be updated; ease of modification is a very important feature in these situations.

We have described the core components of a knowledge-based system: knowledge base, inference engine, user interface, knowledge acquisition tool, and explanation tool. The most important component of a knowledge-based system is the knowledge base, which contains the relevant domain-knowledge.

Several declarative knowledge representation schemas are commonly used; our knowledge base was realised as *rule-based / object-oriented* approach in CLIPS.

As we have mentioned in Section 2.2 image preprocessing and image enhancement have to be the first step after image capturing. This step is necessary to improve the visual appearance of an image and to convert the image in a form better suited for subsequent processing steps. The choice of suitable image preprocessing and image enhancement algorithms can be done by a knowledge-based decision system on the basis of extracted image features. The description of this system is the content of the next chapter.

Chapter 5

Knowledge-Based Image Preprocessing and Image Enhancement System

5.1 Basic Concept

A necessary precondition for the successful application of image processing, like the application of algorithms for finding *interesting points*, is the "quality" of the image. It is often necessary to improve the visual appearance of an image. This can be done by image preprocessing and image enhancement processes.

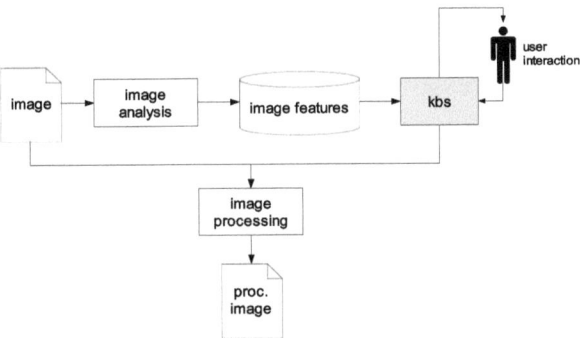

Figure 5.1: Architecture of knowledge-based image preprocessing and image enhancement system.

As we have described in Chapter 3, all decisions are based on the extracted image features. The detailed work flow of the developed knowledge-based image preprocessing and image enhancement system is presented in Figure 5.1. It consists of several steps:

First step – image analysis: After the image is captured by an image sensor (see Section 2.2), the image analysis described in Chapter 3 is carried out. Image features will be stored in a file in a suitable form.

Second step – choice of image processing algorithms: Based on the extracted image features, the knowledge-based decision system chooses an algorithm, a group of algorithms, or a combination of algorithms for image preprocessing and image enhancement in order to improve the image for the subsequent application of *interest operators*. Due to the complexity of these tasks the knowledge-based system works with integrated user-interaction. At critical processing steps the user has the possibility to overrule the decision.

Third step – image processing: The last step in the processing chain is the application of the chosen processing steps. The result is an improved image.

The knowledge-based image preprocessing and image enhancement system consists of two parts (Figure 5.2):

- image preprocessing and image enhancement modules,
- knowledge-based decision system.

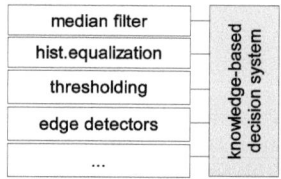

Figure 5.2: Architecture of knowledge-based image preprocessing and image enhancement system.

5.2 Image Preprocessing and Image Enhancement Algorithms

We have classified image processing methods differently into *image enhancement* and *image preprocessing* methods.

Image enhancement methods improve the visual appearance of an image. These methods usually involve the application of image filters to the input image to remove noise and artifacts, smoothen or sharpen the images, or to correct for problems with contrast and/or brightness.

The following image enhancement methods have been implemented:

- histogram equalization,
- grey-level scaling (image brightening / darkening),
- median filtering,
- gauss filtering.

Image preprocessing methods transform the image to a form better suited for the detection of *interest points*. This does not have to be the best form for human analysis. In our case these methods can be used to improve the interpretability (reduction of image information) of an image.

The following image preprocessing methods have been implemented:

- edge detection,
- thresholding.

5.2.1 Histogram Equalization

Histogram equalization redistributes the pixel intensities to equalize their distribution across the intensity range. It may be viewed as an "automatically adjusting contrast filter". The mapping function is given by the cumulative sum of grey-level values:

$$g = \sum_{g'=0}^{g} n(g'), \qquad (5.1)$$

where g is the new grey-level and $n(g')$ is the histogram function of the original image.

The intermediate steps of the histogram equalization process are:

- equalize the cumulative histogram of the image,
- normalize the cumulative histogram to 255,
- use the normalized cumulative histogram as the mapping function of the original image.

The cumulative histogram is a variation of the histogram. The vertical axis does not just give the counts for each grey value, it gives the counts for each grey value plus counts smaller values (Pratt, 1978; Bässmann et al., 1998).

An example of *histogram equalization* is shown in Figure 5.3. The grey value distribution is more uniform compared to that of the original image, so that there is greater contrast in the image.

5.2.2 Grey-Level Scaling (Image Brightening / Darkening)

Grey-level scaling takes an input image and produces an output image in which each pixel value is multiplied by a specified constant. Given a scaling factor greater than one, scaling will brighten an image. A factor less than one will darken the image. Scaling generally produces a much more natural brightening / darkening effect than simply adding an offset to the pixels, since it preserves the relative contrast of the image better (shown in Figure 5.4).

It is important to be aware of what will happen if the multiplications result in pixel values outside the range that can be represented by the image format being used. In our implementation all the pixels which, in the result image, would have a value greater than 255 are wrapped back to 255 (Pratt, 1978).

As we have described above *histogram equalization* and *grey-level scaling* can be used to improve the visual appearance of an image. The resulting images of both methods are nearly the same. For a better understanding of the differences between these two techniques we will compare them in the following.

Comparison between *histogram equalization* and *grey-level scaling*:

- *histogram equalization* is a contrast enhancement technique with the objective to obtain a new enhanced image with an uniform histogram, whereas *grey-level scaling* only multiplies the pixel values by a factor;

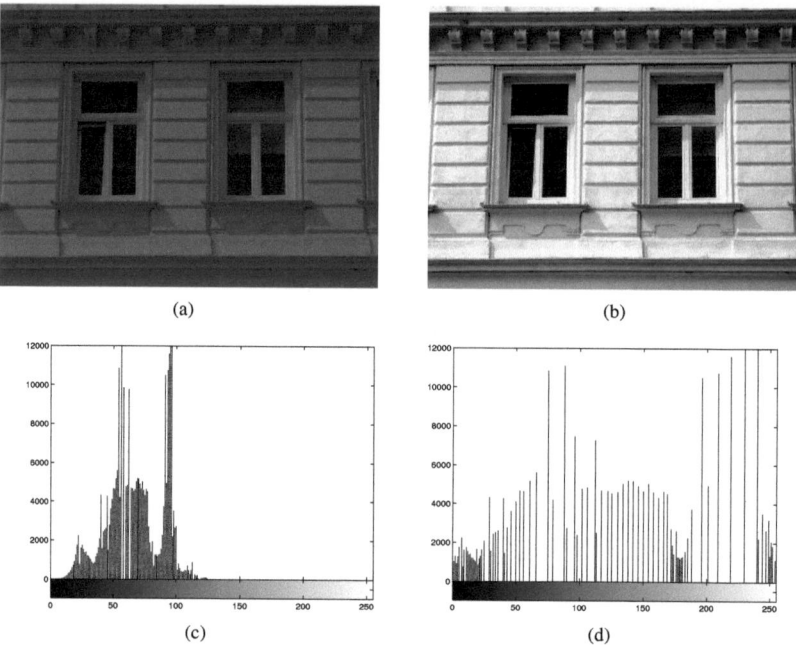

Figure 5.3: (a) A low-contrast image and (b) the image after *histogram equalization*, including their histograms (c, d).

- *histogram equalization* is a method for modifying the dynamic range and contrast of an image, whereas *grey-level scaling* is useful to improve under– or overexposed images and to preserving the relative contrast;

- *histogram equalization* should be used if the histogram has not equally distributed brightness levels over the whole brightness scale, whereas *grey-level scaling* if grey value distribution is uniform but the image under– or overexposed.

5.2.3 Median Filtering

The *median filter* is used to reduce noise in an image. It looks at its nearby neighbours (for example a 3×3 square neighborhood) to decide whether or not it is representative of its surroundings. The *median filter* replaces the neighbouring pixel values with the median of those values. The median is calculated by first sorting all the pixel values from the surrounding neighborhood into numerical order and then replacing the pixel being considered with the middle pixel value (see Figure 5.5).

Median filtering is a simple and very effective noise removal filtering process. Its performance is particularly good for removing shot noise. Shot noise consists of strong spike-like isolated values. Figure 5.6a shows an image that has been corrupted by noise. Applying a 3×3 *median filter* produces an image in which the noise has been entirely eliminated (shown in Figure 5.6).

5.2 Image Preprocessing and Image Enhancement Algorithms

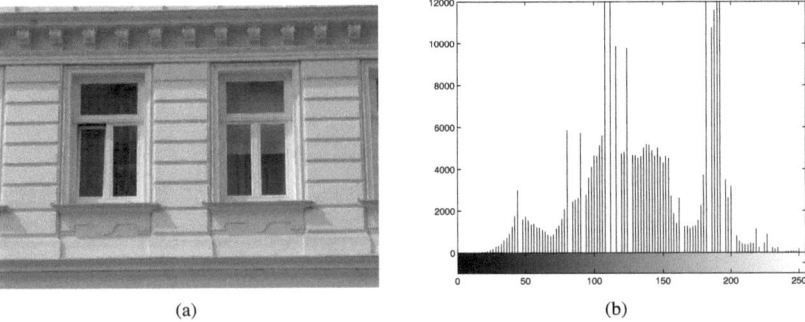

(a)   (b)

Figure 5.4: The example image from Figure 5.3a after *grey-level scaling* by a factor of 2.

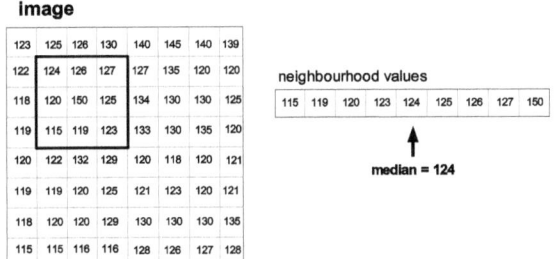

Figure 5.5: Calculating the median value of a 3×3 pixel neighborhood.

5.2.4 Gauss Filtering

The *gauss filter* smoothes or blurs an image by performing a convolution operation with a gaussian filter kernel. The grey value of each pixel is multiplied by the corresponding point in the kernel matrix. The grey-level of the target point is determined by summing all these values for each pixel. We have implemented the *gauss filter* with two fixed kernel sizes:

$$gauss_{3\times 3} = \begin{bmatrix} 1 & 2 & 1 \\ 2 & 4 & 2 \\ 1 & 2 & 1 \end{bmatrix} \qquad gauss_{5\times 5} = \begin{bmatrix} 2 & 7 & 12 & 7 & 2 \\ 7 & 31 & 52 & 31 & 7 \\ 15 & 52 & 127 & 52 & 15 \\ 7 & 31 & 52 & 31 & 7 \\ 2 & 7 & 12 & 7 & 2 \end{bmatrix}$$

The application of *gauss filter* has the effect of defocussing the image. Edges become blurred, narrowed lines become attenuated (see Figure 5.7). The larger the neighborhood of the filter, the greater the defocussing action (Haralick et al., 1993). Often *gauss filtering* is used as preprocessing filter for edge detection methods. We have implemented the *gauss filter* as preprocessing methods for the three edge detection algorithms (see Section 5.2.5).

Median filtering and *gauss filtering* are two smoothing operations which can be used for noise reduc-

Figure 5.6: Noisy image before and after a 3×3 *median filtering*.

Figure 5.7: The noisy image from Figure 5.6a after a 3×3 *gauss filtering*.

tion. For a better understanding which method is suited for which situation, we will discuss some differences between these two filter techniques.

Comparison between *median filtering* and *gauss filtering*:

- *median filtering* is particularly effective at removing random, high-impulse noise from an image;
- *gauss filtering* is very similar to *median filtering*, but it degrades the image lesser (more suited as preprocessing for edge detection);
- *median filtering* produce more edge blurring as *gauss filtering*.

5.2.5 Edge Detection

Edge detection is the process of finding edges in an image. Edges are places in the image with strong intensity contrast. Representing an image by its edges has the advantage that the amount of data is reduced significantly while retaining most of the relevant image information. But even so, edge detection should not be used blindly. We use *edge detection* if the quality of the image is poor and the subsequent application of an *interest operator* would be without success.

5.2 Image Preprocessing and Image Enhancement Algorithms

We have implemented the following *edge detection* algorithms:

- Sobel operator,
- Prewitt operator,
- Laplace operator.

After having described these three *edge detection* methods we will draw a comparison between them.

(a) *Sobel edge detector:* This algorithm implements a relatively simple method to recover extended edge structures, represented as pixel-chains. First a pair of Sobel masks (h_x to detect changes in vertical contrast and h_y to detect horizontal contrast) is applied to the input image. The result of each convolution is treated a vector representing the edge through the current pixel. If the magnitude of the sum of these two orthogonal vectors is greater than a threshold, the pixel is marked as an edge.

The two convolution kernels used are:
$$h_x = \begin{bmatrix} -1 & 0 & 1 \\ -2 & 0 & 2 \\ -1 & 0 & 1 \end{bmatrix} \qquad h_y = \begin{bmatrix} -1 & -2 & -1 \\ 0 & 0 & 0 \\ 1 & 2 & 1 \end{bmatrix}$$

The *Sobel detector* is incredibly sensitive to noise in the picture. The detected edges are also quite thick. This can occur if the edge is one pixel thick – the *Sobel detector* will thicken it because there are two consecutive intensity changes (Pratt, 1978). For this reason the *Sobel edge detector* is ineligible for images with capillary patterns, like typical old building facades.

An example of edge detection with the *Sobel operator* is presented in Figure 5.8.

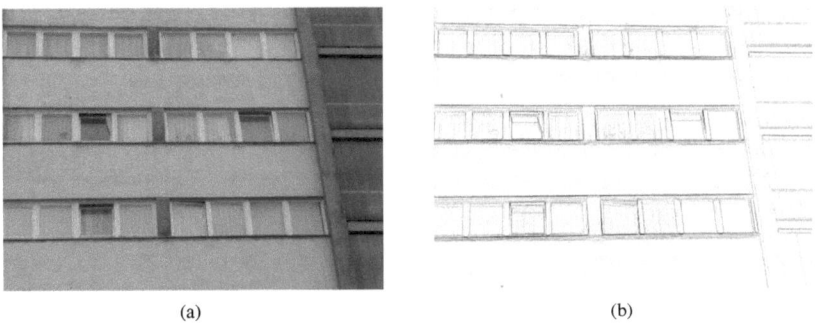

(a)   (b)

Figure 5.8: Example of Sobel edge detection.

(b) *Prewitt edge detector:* This algorithm is one of the oldest and best understood methods of detecting edges in images. As the *Sobel operator* this algorithm works with two convolution masks, one for detecting image derivatives in horizontal direction and one for detecting image derivatives in vertical direction. So it provides *directional edge detection*. The main problem of the *Prewitt edge detector* is the possibly existing noise in images. The spikes in the derivative from the noise can mask the real maxima that indicate edges. If we smooth the image then the effects of noise can be reduced.

The two convolution kernels used are:

$$h_x = \begin{bmatrix} -1 & 0 & 1 \\ -1 & 0 & 1 \\ -1 & 0 & 1 \end{bmatrix} \quad h_y = \begin{bmatrix} -1 & -1 & -1 \\ 0 & 0 & 0 \\ 1 & 1 & 1 \end{bmatrix}$$

An example of edge detection with the *Prewitt operator* is presented in Figure 5.9.

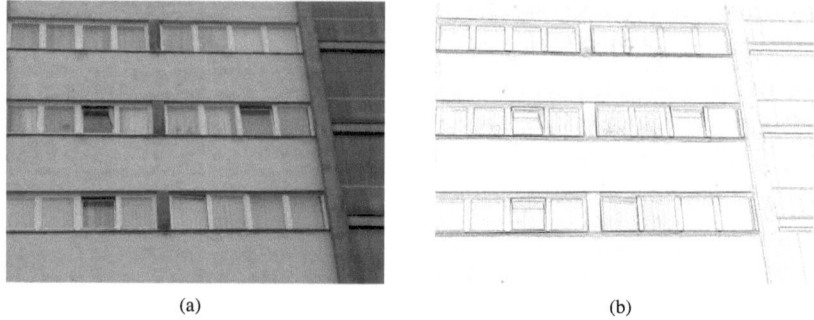

(a)   (b)

Figure 5.9: Example of Prewitt edge detection.

(c) *Laplace edge detector:* This algorithm is an isotropic measure of the 2nd spatial derivative of an image. The *Laplace edge detector* highlights regions of rapid intensity change and is therefore often used for edge detection. The operator normally takes a single grey-level image as input and produces another grey-level image as output. Since the input image is represented as a set of discrete pixels, the second derivatives in the definition of the Laplacian can be approximated with convolution kernels.

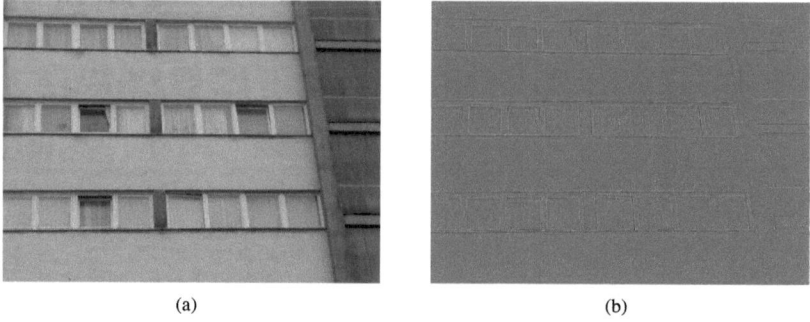

(a)   (b)

Figure 5.10: Example of Laplace edge detection.

Like the *Sobel* and the *Prewitt edge detector*, one drawback of this operator is that it is very sensitive to noise. That is why it is necessary to filter the image before applying edge detection. The edges detected by the *Laplace edge detector* are very weak. Often the *Laplace operator*

5.2 Image Preprocessing and Image Enhancement Algorithms

is combined with a *Gaussian smoothing filter* (see Section 5.2.4). The filter is then called *Laplacian of Gaussian* (LoG).

The two convolution kernels used are:

$$h_1 = \begin{bmatrix} 0 & -1 & 0 \\ -1 & 4 & -1 \\ 0 & -1 & 0 \end{bmatrix} \qquad h_2 = \begin{bmatrix} -1 & -1 & -1 \\ -1 & 9 & -1 \\ -1 & -1 & -1 \end{bmatrix}$$

An example of edge detection with the *Laplace operator* is presented in Figure 5.10.

Having described the three implemented edge detection algorithms we will compare the different techniques. This is necessary to understand the usability of the different methods.

Comparison between the implemented edge detection methods:

- the result of the *Sobel Operator* (combined h_x and h_y filter) is almost independent from the orientation of the edge it detects;
- the *Sobel operator* is less sensitive to isolated high-intensity point variations since the local averaging over sets of three pixels tends to reduce this;
- the *Sobel operator* still produced very thick edges (image with smoother, more pronounced outlines) that have to be thinned by morphological operations (e.g *erosion* or *dilation*);
- the *Prewitt Operator* is more sensitive to the edge orientation than the *Sobel operator*;
- the *Laplace operator* is the most sensitive edge detector with regard to noise;
- when applying the *Laplace operator* to the images, many "gaps" are obtained in the edges (see Figure 5.10b);
- the *Laplace operator* produces an image with harsh intensity transitions, and results in an image with only edges of high contrast visible.

More details about edge detection can be found in (Pratt, 1978; Haralick et al., 1993).

5.2.6 Thresholding

Thresholding is a technique for converting a grey-scale or color image to a binary image based upon a threshold value (*th*).

If a pixel in the image has an intensity less than the threshold value, the corresponding pixel in the resultant image is set to white. Otherwise, if the pixel is greater than or equal to the threshold intensity, the resulting pixel is set to black.

It is possible to separate (to segment) different parts of an image on the basis of pixel intensity. In this case, we expect to see a distinct peak in the histogram corresponding to the parts so that thresholds can be chosen to isolate this peak accordingly (Figure 5.11). Traditional *thresholding* has been based on suppressing background rather than finding foreground. It can also be used as a postprocessing filter for edge detection. Noise that is still present in the output image (after edge detection) can be reduced (Figure 5.12). In our system *thresholding* is used as a postprocessing filter for the three implemented edge detection methods (see Section 5.2.5).

40   Knowledge-Based Image Preprocessing and Image Enhancement System

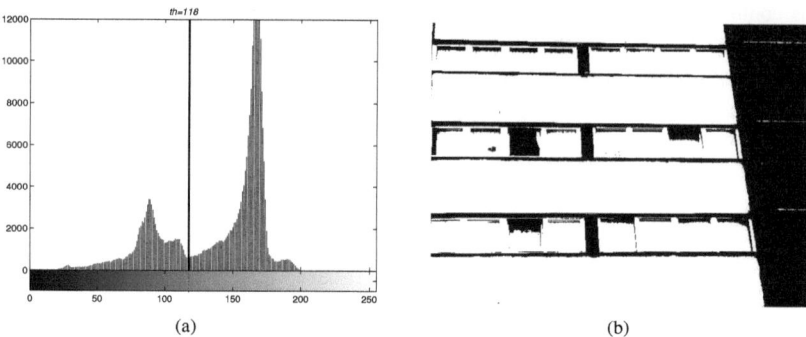

Figure 5.11: *Thresholding* ($th = 118$) of Figure 5.8a including the histogram of the original image with suitable choices of threshold.

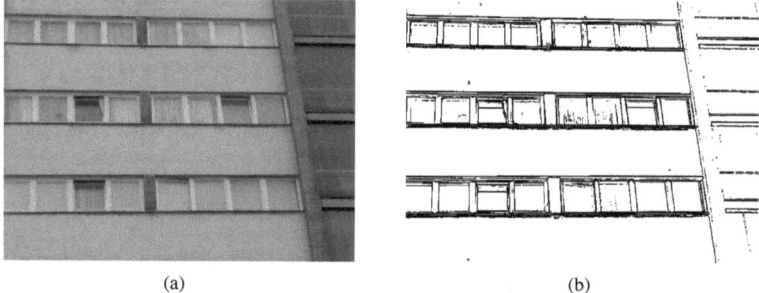

Figure 5.12: Example of *thresholding* ($th = 64$) after a *Prewitt edge detector*.

5.2.7 Overview

Figure 5.13 gives a graphical overview of the described image enhancement and image preprocessing methods.

All algorithms for a image enhancement and image preprocessing system have been described in the previous sections. For an automatic decision system the knowledge about the usability, ability and the context between extracted image features and algorithms has to be implemented in the knowledge base. The description of this implementation will be the content of the next section.

5.3 The Knowledge Base – Image Preprocessing and Image Enhancement

As we explained in Section 4.2.1 the knowledge base is one of the core components of a knowledge-based system. It contains the relevant domain-knowledge that the knowledge engineer has implemented in the course of the development process.

5.3 The Knowledge Base – Image Preprocessing and Image Enhancement

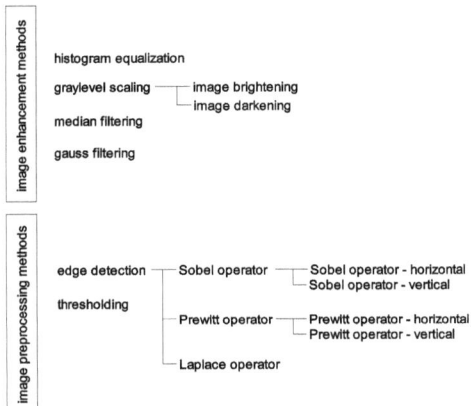

Figure 5.13: Overview of the implemented image enhancement and image preprocessing methods.

The knowledge which was required to be included in this part of the knowledge base was obtained in different ways: from technical literature (Pratt, 1978; Bässmann et al., 1998), previous projects (Roic, 1996; Mischke, 1998) and from experiments.

The knowledge base (the part for the knowledge-based image preprocessing and image enhancement system) is divided into three groups of rules:

- rules for the choice of suitable algorithms for image preprocessing and image enhancement,
- rules for the predefinition of necessary parameters,
- rules to define the order of the algorithms.

In the following we will discuss each of these rule groups including the necessary preconditions of the implemented image preprocessing and image enhancement steps.

In order not to anticipate the concrete implementation of the rules and their syntax, we will list the preconditions in form of tables. A listing of the complete knowledge base can be found in Appendix C.1. The syntax of a rule and its relevant concept has been described in Section 4.2.1.

5.3.1 Rules for the Choice of Suitable Algorithms

(a) *Histogram equalization* is a useful method for modifying the dynamic range and contrast of an image (see Section 5.2.1). The necessary preconditions for a histogram equalization are listed in Table 5.1. This image enhancement algorithm is not dependent on other methods.

(b) *Grey-level scaling* (image brightening / darkening) is useful to improve under- or overexposed images. It produces a much more natural brightening / darkening effect, since it preserves the relative contrast of the image. We have divided this method into two submethods: image brightening and image darkening (see Section 5.2.2). The necessary preconditions for both

Histogram equalization	
M_1	very low *or* low *or* very high *or* high
M_3	low positive *or* very low positive *or* low negative *or* very low negative

Table 5.1: Preconditions for histogram equalization.

submethods are listed in Table 5.2 and 5.3. This image enhancement algorithm is not dependent on other methods.

Image brightening	
M_1	very low *or* low
M_3	middle negative *or* high negative *or* very high negative *or* middle positive *or* high positive *or* very high positive

Table 5.2: Preconditions for grey-level scaling (image brightening).

Image darkening	
M_1	very high *or* high
M_3	middle negative *or* high negative *or* very high negative *or* middle positive *or* high positive *or* very high positive

Table 5.3: Preconditions for grey-level scaling (image darkening).

(c) *Median filtering* is an image enhancement method for local smoothing by replacing each pixel with the median grey-level of neighboring pixels (see Section 5.2.3). This method has no necessary preconditions. Its application depends only on the user interaction.

(d) *Gauss filtering* smoothes the image and reduces image noise. The application of a *gauss filter* depends on the application of edge detection algorithms. If edge detection is necessary *gauss filtering* is done beforehand (see Section 5.2.4).

(e) *Edge detection* is the process of finding places in the image with strong intensity contrast (see Section 5.2.5). Choosing edge detection is a two step process:

- verification if edge detection is necessary,
- choosing a suitable edge detection algorithm.

As we have explained, we use *edge detection* only if the image has poor quality and the subsequent application of an *interest operator* would be without success. The preconditions for the verification of necessary edge detection are listed in Table 5.4.

After having decided that edge detection is necessary, a suitable edge detection algorithm is selected from the implemented one (*Sobel operator*, *Prewitt operator* or *Laplace operator*). As we have explained in Section 5.2.5 the *Sobel operator* and the *Prewitt operator* are able to detect edges in vertical direction separated from edges in horizontal direction. Therefore these two edge detection methods have an additional precondition, which relies on the homogeneity in these two directions. These preconditions are listed in Table 5.5.

The preconditions for the three edge detection methods are listed in Table 5.6, 5.7 and 5.8.

5.3 The Knowledge Base – Image Preprocessing and Image Enhancement

Edge detection	
M_1	very low
M_2	very low
M_3	middle negative *or* high negative *or* very high negative *or* middle positive *or* high positive *or* very high positive

Table 5.4: Preconditions for edge detection.

	Horizontal homogeneity	Vertical homogeneity
$H_1 - 0°$	very high *or* high	not very high *or* not high
$H_5 - 0°$	very high *or* high	not very high *or* not high
$H_1 - 90°$	not very high *or* not high	very high *or* high
$H_5 - 90°$	not very high *or* not high	very high *or* high

Table 5.5: Preconditions for horizontal and vertical homogeneity.

(f) *Thresholding* is a segmentation technique with which it is possible to separate out different parts of an image on the basis of pixel intensity (see Section 5.2.6). In our system this method is used as post-processing filter for the three edge detection methods implemented. Therefore the *thresholding* process only has the precondition that edge detection has been selected.

After having chosen the image preprocessing and image enhancement methods it is necessary to appoint their parameters. This process will be described in the next section.

5.3.2 Rules for the Predefinition of Necessary Parameters

(a) *Histogram equalization* needs no parameters.

(b) *Grey-level scaling* needs the *scaling factor* as a parameter. A *grey-level scaling* with a scaling factor greater than one will brighten an image; a factor less than one will darken the image. The goal of *grey-level scaling* is to obtain a much more natural brightening / darkening effect. The scaling factor can be chosen in such a way that the histogram feature *mean* (M_1) (see Equation 3.2) after the *grey-level scaling* has a value of 128.

Thus in our system *grey-level scaling factor* is defined as:

$$factor_{gs} = \frac{128}{M_1}. \tag{5.2}$$

(c) *Median filtering* needs the length of the filtering window (see Section 5.2.3) as its parameter. In our system this parameter has to be specified by the user. The default value is 3.

(d) *Gauss filtering* has the kernel size as its parameter. As we have explained in Section 5.2.4 our implementation contains two fixed kernel sizes: the $gauss_{3\times 3}$ and the $gauss_{5\times 5}$ kernel. The choice has to be made by the user. The first one is the default kernel.

(e) The implemented *edge detection* algorithms (Sobel-, Prewitt- and Laplace operator) need no parameters.

Sobel edge detection – horizontal		Sobel edge detection – vertical	
horizontal homogeneity	yes	vertical homogeneity	yes
H_9 – $average$	very low or low	H_9 – $average$	very low or low

Table 5.6: Preconditions for the *Sobel edge detection*.

Prewitt edge detection – horizontal		Prewitt edge detection – vertical	
horizontal homogeneity	yes	vertical homogeneity	yes
H_9 – $average$	middle	H_9 – $average$	middle

Table 5.7: Preconditions for the *Prewitt edge detection*.

(f) The parameter for *thresholding* is the so-called threshold value *th* (see Section 5.2.6). In our system *thresholding* is used as a post-processing filter for the implemented edge detection methods. For this special case the threshold value can be equated to the histogram feature *mean* (M_1) (see Equation 3.2):

$$th = M_1. \tag{5.3}$$

5.3.3 Rules to Define the Order of the Algorithms

These rules go through the list of chosen algorithms and decide which actions have to be carried out in which order.

(a) *Histogram equalization* or *grey-level scaling* is always the first step in the image preprocessing and image enhancement process.

(b) *Median filtering* is the first step or the first step after *histogram equalization / grey-level scaling*;

(c) *Gauss filtering* is always the step before *edge detection*;

(d) *Edge detection* is always the step after *gauss filtering* and before *thresholding*;

(e) *Thresholding* is the first step after *edge detection*.

5.3.4 Overview – Rule Base

Due to the small number of implemented image preprocessing and image enhancement algorithms, the knowledge base could be kept propositionally simple and thus easily modifiable and extensible.

Before we illustrate two examples we will give some additional information about the developed knowledge base.

The complete knowledge base for the choice of suitable image preprocessing and image enhancement algorithms consists of 37 rules. The three "sub rule bases" are:

- rules for the choice of suitable algorithms – 15 rules;
- rules for the predefinition of necessary parameters – 5 rules;
- rules to define the order of the algorithms – 9 rules.

5.4 Examples

Laplace edge detection	
H_9 − $average$	high or very high

or

Laplace edge detection	
Edge detection	yes
Sobel operator	no
Prewitt operator	no

Table 5.8: Preconditions for the *Laplace edge detection*.

Additionally the knowledge base contains rules to display and write the necessary output, rules for user-interaction and rules to document conclusions (a total of 8 rules). A listing of the complete knowledge base can be found in Appendix C.1.

5.4 Examples

In this section we will provide a better understanding of the functionality of the implemented knowledge-based image preprocessing and image enhancement system by means of two examples.

5.4.1 First Example:

This first example shows an noisy underexposed image (Figure 5.14a). The histogram (Figure 5.14b) shows that only a few grey-level values are present. Most of the grey values show zero values in the histogram, indicating that none of the pixels have those values. Some details in this underexposed image are not or not easily visible. To bring out useful image details suitable image preprocessing and image enhancement algorithms may be used.

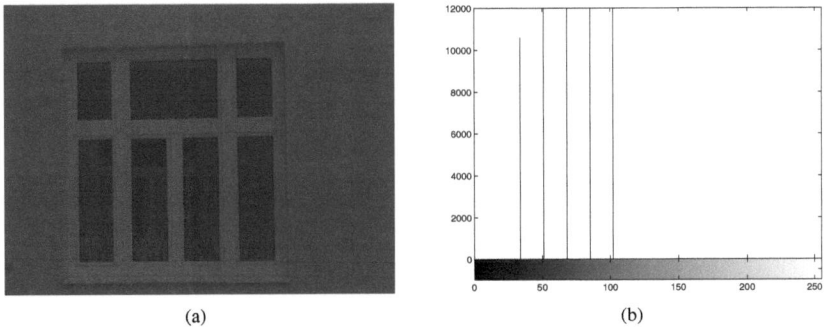

(a)   (b)

Figure 5.14: Noisy underexposed image.

The extracted image features are listed in Table 5.9.

If we compare the preconditions of the described rules (Section 5.3) with the extracted image features from Table 5.9 the following image preprocessing and image enhancement algorithms have to be done: *grey-level scaling* (*image brightening*), *median filtering* and *edge detection*.

M_1	M_1 - f.v.	M_2	M_2 - f.v.	M_3	M_3 - f.v.				
69.8453	very low	11.6399	very low	-0.9996	very high negative				
	0°	f.v.	45°	f.v.	90°	f.v.	135°	f.v.	average	f.v.

	0°	f.v.	45°	f.v.	90°	f.v.	135°	f.v.	average	f.v.
H_1	0.0213	v.high	0.0165	v.high	0.0189	v.high	0.0167	v.high	0.0184	v.high
H_2	1.0110	v.low	0.4241	v.low	0.8906	v.low	0.4362	v.low	0.6905	v.low
H_5	0.7375	v.high	0.6523	v.high	0.7004	v.high	0.6560	v.high	0.6866	v.high
H_9	1.9333	v.low	2.0604	low	1.9825	v.low	2.0553	low	2.0079	low

Table 5.9: Extracted image features for Figure 5.14a.

As *edge detection* is selected, the knowledge-based system gives the user the possibility to overrule this decision. If the user makes use of this, *edge detection* will not be applied; *grey-level scaling* (*image brightening*) and *median filtering* will remain. The processed image is shown in Figure 5.15a.

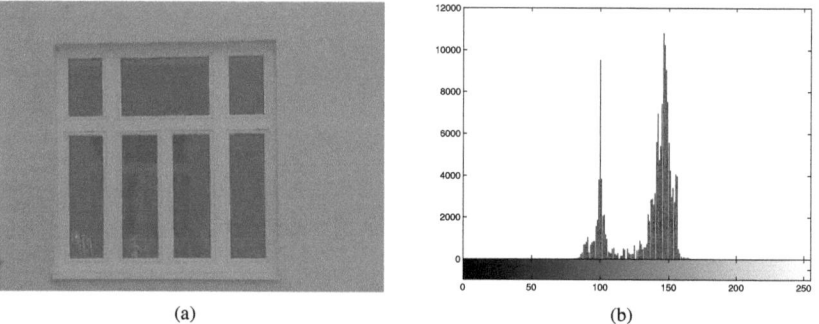

(a)　　　　　　　　　　　　　　　　(b)

Figure 5.15: The noisy underexposed image from Figure 5.14 after *image brightening* and a 3×3 *median filtering*.

The reassignment of grey values has increased the visual contrast for the present pixels; therefore the histogram of the resulting image (Figure 5.15b) is more spread. The application of the *median filter* has reduced the impulse noise by smoothing the image. Useful details, like edges and corners, are now, after image preprocessing and image enhancement, visible.

If the user does not overrule the system decision *edge detection* has to be carried out. As we have described in Section 5.3, *edge detection* has *gauss filtering* as a preprocessing method (which replaces *median filtering*) and *thresholding* as a postprocessing method. It follows the new process sequence: *gauss filtering*, *histogram equalization*, *edge detection (Laplace operator)* and *thresholding*. The resulting image is shown in Figure 5.16.

The applied edge detection (*Laplace edge detection*) and *thresholding* have reduced the information content of the image substantially. The resulting image is completely different from the image shown in Figure 5.15. Grey-level values have been lost and only places with strong intensity contrast have been conserved. First of all edges and corners of the windows have been extracted. Additionally the rough masonry results in an immense number of small edge particles. For this fact the image preprocessing and image enhancement sequence without edge detection is more suited for subsequent processing steps, like finding *interesting points*.

Figure 5.16: The noisy underexposed image from Figure 5.14 after a 3×3 *gauss filtering*, *image brightening*, 3×3 *Laplace edge detection* and *thresholding*.

5.4.2 Second Example:

For the second example we have chosen a highly overexposed noisy image (Figure 5.17a). Specifically the noise in real-world occurs in all sorts of ways. Capturing images without any noise is impossible. Among its most frequent causes are instability in the light source or detector during the time required to capture the image, scanning errors, compression artifacts and others (Pratt, 1978). Our example (a rather extreme example) contains strong regular impulse noise. The histogram (5.17b) shows that the image has low contrast (narrow peak) and most of the grey values are too bright (values greater than 120).

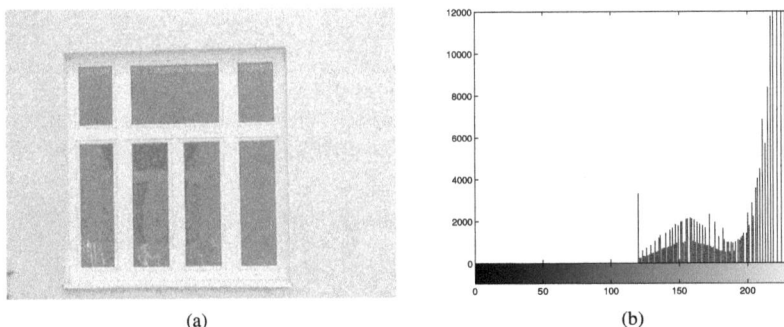

(a)   (b)

Figure 5.17: The noisy overexposed image.

Due to the calculated image features (Table 5.10) the knowledge-based system has chosen the following image preprocessing and image enhancement algorithms: *median filtering* and *grey-level scaling* (*image darkening*). *Image darkening* is needful to correct the brightness of the image. Such a expansion of contrast is mostly accompanied by an increase of visibility for noise that may be present. Because of this fact, median filtering is advisable. The resulting image is shown in Figure 5.18.

M_1	M_1 - f.v.	M_2	M_2 - f.v.	M_3	M_3 - f.v.				
209.5469	very high	34.2994	middle	-0.9383	very high negative				
	0°	f.v.	45°	f.v.	90°	f.v.	135°	f.v.	average	f.v.
H_1	0.0056	v.high	0.0056	v.high	0.0056	v.high	0.0056	v.high	0.0056	v.high
H_2	189.889	mid.	200.082	high	168.217	mid.	200.075	high	189.566	mid.
H_5	0.2054	low	0.2035	low	0.2066	low	0.2041	v.low	0.2049	low
H_9	2.8900	v.low	2.8978	low	2.8833	low	2.8978	low	2.8922	low

Table 5.10: Extracted image features for Figure 5.17a.

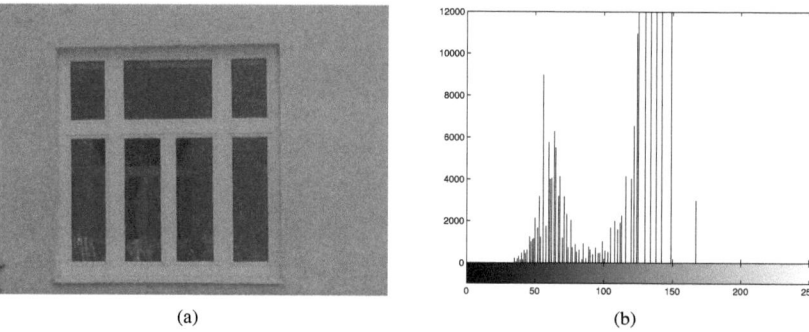

(a)   (b)

Figure 5.18: The image from Figure 5.17 after a 3×3 *median filtering* and *grey-level scaling*.

5.5 Summary

In this chapter we have presented a knowledge-based method for image preprocessing and image enhancement. The choice of such algorithms is based on extracted image features (histogram features and Haralick features) calculated by a specially developed image analysis (see Chapter 3). Due to the complexity of this task the knowledge-based system works with integrated user-interaction. At critical processing steps (e.g. *median filtering* and *edge detection*) the user has the possibility to overrule the decision.

The developed knowledge-based image preprocessing and image enhancement method does not have to be seen as a stand-alone process, but in the context of the global procedure of measurement described in Chapter 2.

The developed knowledge-based image preprocessing and image enhancement system was tested on about 120 images. It yields suitable results for the subsequent process steps and shows a reasonable performance for an on-line measurement system (less than 2 seconds for a picture including image analysis). The runtime of image preprocessing and image enhancement methods has no effect on the performance (runtime) of the whole measurement process.

After having prepared the image the next sequence step can follow. As we have described in Section 2.2, after image preprocessing and image enhancement the point finding process can be started. This process will be the content of the next chapter.

Chapter 6

Knowledge-Based Point Detection

6.1 Basic Concept

After having improved the visual appearance of an image by image preprocessing and image enhancement processes, point finding in the image can follow. Processing algorithms which extract *interesting points* are called *interest operators* (IOPs). *Interest operators* play an important role in computer vision and image processing. They highlight points which can be found easily by using correlation methods. There is a huge number of *interest operators* (Moravec, 1977; Förstner, 1991; Harris et al., 1988; Paar et al., 2001), however no *interest operator* is suitable for all types of desired point detection. For this reason we have implemented different *interest operator* algorithms in our system. The choice of a suitable algorithm and its parameter is made by the knowledge-based system, which is described in this chapter. Thereto we expand the architecture of our system shown in Figure 5.1 by the knowledge-based point detection system (Figure 6.1).

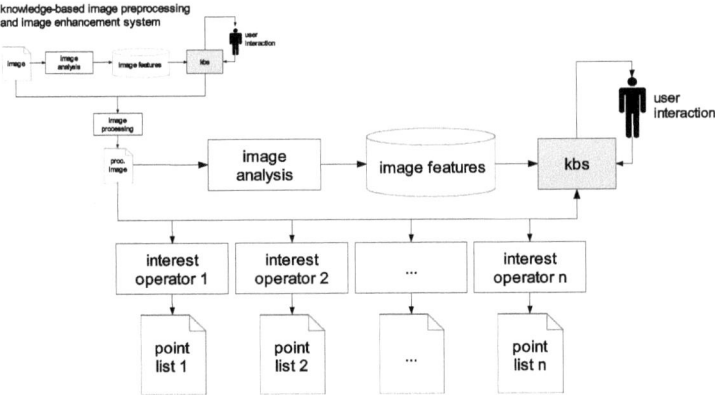

Figure 6.1: Architecture of knowledge-based point detection system (including the preceding processing steps).

Basically the system for the knowledge-based point detection (right part of Figure 6.1) consists of two main components:

- point detectors (*interest operators*),
- knowledge-based system.

At the beginning of the process chain image analysis of the processed images (done by the image preprocessing and image enhancement system) has to be performed, since the image properties have been changed. Based on these extracted image features (Chapter 3), the knowledge-based system chooses a suitable algorithm (*interest operator*) for finding interest points.

As we have explained in Section 1.1 the application of the developed measurement system is focused on monitoring of building facades. For such a process the facade has to be modeled by choosing points in such a way that they characterize the object. In a subsequent process step these points can be used for object reconstruction or monitoring of movements and distortions. Building facades elements (e.g. edges, windows, beams) can be represented by simple line geometry. Therefore the process of object modeling can be reduced on capturing points along such lines respectively intersections of them. The choice of suitable *interest operators* by the knowledge-based system is fitted according to this.

As we have explained above no *interest operator* is suitable for all types of desired point detection (i.e. for all types of line geometry). For this reason we have implemented three different *interest operators* into our system. Not only the best *interest operator* will be selected, but also a group of suitable algorithms including their ranking. In the course of finding *interest points* one or more *interest operators* will be applied.

The following *interest operators* have been implemented:

- Förstner operator (Förstner, 1991),
- Harris operator (Harris et al., 1988),
- Hierarchical Feature Vector Matching operator (Paar et al., 2001).

More details about these algorithms and their functionality in the next section.

6.2 Interest Operators

By *"interesting points"* we mean any point in the image for which the signal (the grey values of image pixels) changes two-dimensionally. Image processing methods which locate striking points are called *"interest operators"*.

Förstner (1991) gives the following definition for *"interest operator"*:

> *A fast operator for detection and precise location of distinct points, corners and centres of circular features.*

In literature there are a wide variety of *interest operators*. Schmid et al. (2000) differentiate three categories:

(a) *Contour based methods* extract contours in the image and search for maxima curvature or inflexion points along the contour chains.

6.2 Interest Operators

(b) *Intensity based methods* compute measurements directly from grey values that indicate the presence of *interesting points*.

(c) *Parametric model based methods* fit a parametric intensity model to the signal.

The algorithms implemented in our system are intensity based methods. These methods go back to the development done by Moravec (1977). His detector is based on the auto-correlation function of the signal. It measures the grey value differences between a window and a window shifted in the four directions parallel to the row and columns. An interest point is detected if the minimum of these four directions is superior to a threshold (Schmid et al., 2000). Today there are different improvements and derivatives of the *Moravec* operator. Among the most well-known are the Förstner and the Harris operator, which represent two methods implemented into our system. Additionally we have integrated the *Hierarchical Feature Vector Matching (HFVM) operator*, a development of the *Joanneum Research* in Graz (Austria).

6.2.1 Förstner Operator

The Förstner operator is one of the most famous and popular *interest operators* in computer vision. It was developed more than 20 years ago by *Wolfgang Förstner*. It is based on the assumption that a corner point is the point that is statistically closest to all the edge elements along the edges intersecting at that corner (the point location is determined through a least squares adjustment procedure). The operator evaluates the quality of the corner points by analyzing the shape and the size of error ellipses describing the variance covariance matrix associated with the derived corner point location.

In order to quantify these properties, the parameters "roundness" q and "size" W of the error ellipse are defined:

$$q = \frac{4 \cdot det N}{(trace N)^2}, \quad W = \frac{det N}{trace N}, \quad (6.1)$$

$$N^{-1} = \begin{bmatrix} gu^2 & gu \cdot gv \\ gu \cdot gv & gv^2 \end{bmatrix}, \quad (6.2)$$

where gu and gv are the derivatives of the grey values of image pixels across the rows (u) and columns (v) of the window. An interest point is defined by values of W and q greater than some thresholds (W_{min} and q_{min}) and extreme maximum values in the neighborhood.

Reliable corner points should have a near circular error ellipse with a small size. A larger W indicates a smaller error ellipse and a circular error ellipse will have a maximum q value of 1.

To sum up it can be said that the Förstner operator works in a two step process:

- detecting points by searching for optimal windows (windows of length R) using the auto-correlation matrix;
- improvement of the accuracy within the detected windows.

Analyzing the local gradient field the Förstner operator is able to classify *interesting points* into junctions or circular features.

A listing of the complete mathematical derivation can be found in (Förstner, 1991; Mischke, 1998).

We use two different implementations of the Förstner operator. The first one (in the following we call it Förstner 1) is based on the implementation of the original author (Förstner, 1991). It works with fixed parameters for q_{min} and W_{min}. Only the window length R can be chosen by the user or by the system. The second one (Förstner 2) is an improved version of the original implementation and was developed by Mischke (1998). This implementation permits the selection of values for q_{min}, W_{min} and R.

Two example images with detected points by the Förstner 2 operator are shown in Figure 6.2. On the left picture (Figure 6.2a) the Förstner 2 operator was executed with $q_{min} = 0, 2$; $W_{min} = 400$ and $R = 3$; on Figure 6.2b with $q_{min} = 0, 2$; $W_{min} = 450$ and $R = 3$.

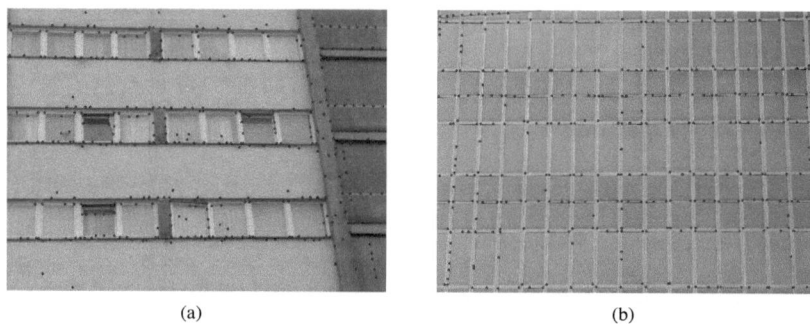

(a)   (b)

Figure 6.2: Extracted points by the Förstner 2 operator.

In Figure 6.2, 369 and 343 points were detected in the left and right image respectively. We can observe that the Förstner operator extracts mainly points on striking edges and line intersections. Only a small number of isolated single points are detected. It is notable that the operator extracts many *interest points* on the right part of Figure 6.2a too, in spite of low contrast.

A detailed comparison and evaluation of the three implemented *interest operators* was carried out in the course of the knowledge engineering process and will be described in Section 6.3.1.

6.2.2 Harris Operator

The Harris operator, a development by *Chris Harris*, is an improvement of the *Moravec* approach. Instead of using a simple sum, a gaussian is used to weight the derivatives inside the window. *Interest points* are detected if the auto-correlation matrix has two significant eigenvalues.

The Harris operator consists of the following process steps (Harris et al., 1988):

- smooth the image by convolving it with a gauss filter $G(x, y)$;
- compute the image gradient $\nabla I(x, y)$ for each pixel:

$$\nabla I(x,y) = \left[\frac{\partial I(x,y)}{\partial x} , \frac{\partial I(x,y)}{\partial y} \right] ; \qquad (6.3)$$

6.2 Interest Operators

- compute the symmetric positive semi-definite 2×2 matrix A for each pixel and a given size of N_0 (the integrative scale σ_I) as follows:

$$A = \sum_{(x,y) \in N_0} \nabla I(x,y) \nabla I(x,y)^T ; \qquad (6.4)$$

- evaluate the response function for each pixel $R(x, y)$:

$$R(x,y) = corn = det A - \kappa \, trace^2 A , \qquad (6.5)$$

where $\kappa = 0.04$;

- choose the interest point as local maximum of function $R(x, y)$.

Our implementation of the Harris operator works according to this scheme and needs three input parameters: the standard deviation σ as measured for the level of image smoothing (called *derivative scale*), the size of the neighborhood N_0 (called *integration scale σ_I*), and the threshold for the response function ($corn_{min}$).

(a)   (b)

Figure 6.3: Extracted points by the Harris operator.

A listing of the complete mathematical derivation can be found in (Harris et al., 1988). Figure 6.3 shows two picture examples with points detected by the Harris operator. The left picture (Figure 6.3a) was detected with the input parameter $\sigma = 1.0$, $\sigma_I = -0.04$ and $corn_{min} = 0.0035$; the right image (Figure 6.3b) with $\sigma = 1.0$, $\sigma_I = -0.04$ and $corn_{min} = 0.0053$.

In Figure 6.3 we can observe that the Harris operator with the chosen input parameters detects more isolated single points in comparison with the Förstner operator (Figure 6.2). However, the Harris operator extracts more points on edges than the Förstner operator.

6.2.3 Hierarchical Feature Vector Matching (HFVM)

Hierarchical Feature Vector Matching (HFVM) was developed at the Institute of Digital Image Processing of *Joanneum Research* in Graz (Austria), as a new matching technique (Paar et al., 2001). Part of the whole process is the detection of *interesting points*. This part is used by us and will be called in the following text in a simplified way as Hierarchical Feature Vector Matching operator.

Many matching techniques use one or at most two different properties of an image (e.g. grey-levels or/and corners). To improve these techniques, different features should be combined. A feature describes the neighborhood of a pixel.

The HFVM operator is based on the idea of creating a feature vector for each pixel in the image (it results in a so-called feature image). This feature vector contains all the features of one location for the corresponding pixel. Finding a match means comparing a feature vector of a reference image (the so-called reference vector), with all feature vectors of the search area which is part of the search image. The robustness and efficiency of the algorithm will be improved by creating pyramids of the input image (pyramid levels). Matching starts from level 0 (original image). The result of each pyramid level, the so-called disparity map, is used as input for the matching of the next level.

Paar et al. (2001) have formulated the major steps of the HFVM operator as follows:

- build the image pyramid;
- compute the feature images for each pyramid level;
- match the top level of the pyramid;
- filter the resulting disparity map;
- check matching consistency by backmatching;
- interpolate the undefined disparities;
- use the resulting disparity map as the initial disparity map to match the next lower pyramid level.

Due to the possibility to choose and combine any features, the HFVM operator is a flexible algorithm which can be used for many types of applications, such as 3D surface reconstruction (Paar et al., 1996), navigation of robots (Kolesnik et al., 1998), and, as described here, for the detection of *interest points*.

The HFVM operator needs as input a feature set, e.g. horizontal high pass, vertical high pass, horizontal band pass, vertical band pass, and gaussian (Paar et al., 1996).

In our implementation we currently use only the following feature (suited for finding line-intersections):

$$feature_{HFVM} = \begin{bmatrix} 1 & 1 & 0 & -1 & -1 \\ 1 & 1 & 0 & -1 & -1 \\ 0 & 0 & 0 & 0 & 0 \\ -1 & -1 & 0 & 1 & 1 \\ -1 & -1 & 0 & 1 & 1 \end{bmatrix} \quad (6.6)$$

The main problem upon application of the HFVM operator is choosing an appropriate feature set. If the feature set is not suitable, point detection is inadequate – Figure 6.4a shows such a case; whereas the result shown in Figure 6.4b is much better.

The whole functionality and a detailed description of the Hierarchical Feature Vector Matching operator can be found in (Paar et al., 2001).

All *interest operators* have been described in the previous sections. For an automatic decision system, knowledge about these processes and their applicableness has to be implemented in the knowledge

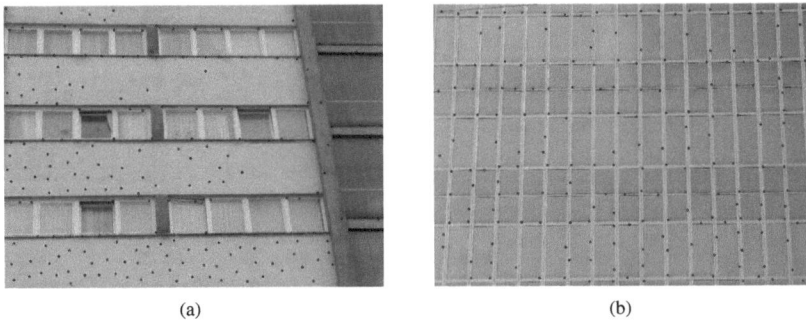

Figure 6.4: Extracted points by the Hierarchical Feature Vector Matching operator.

base. This implementation including the process of knowledge collecting (knowledge engineering) will be the content of the next section.

6.3 The Knowledge Base – Point Detection

As we explained in Section 4.2.1, the knowledge base contains the relevant domain-knowledge that the knowledge engineer has implemented in the course of the development process.

For a clear structure the knowledge base (the part for the knowledge-based point detection) is divided, similarly to the part described in Section 5.3, into three groups of rules:

- rules for the choice of a suitable interest operator,
- rules for the predefinition of necessary parameters,
- rules to define the order of the algorithms.

In the following, we will discuss each of these rule groups including the necessary preconditions of the implemented *interest operators*.

The knowledge, to be included in this part of the knowledge base, was obtained by extensive knowledge engineering. Only a few authors have evaluated and compared *interest operators* (Baker et al., 1999; Bowyer et al., 1999), mentionable is the work done by Schmid et al. (2000).

The problem of existing "interest operator comparing methods" is that most of them are based on human visual inspection or ground-truth verification. These rely on human symbolic interpretation of the image and are therefore subjective. Furthermore, human interpretation limits the complexity of the image used for evaluation (Schmid et al., 2000).

For these reasons we use several methods for the evaluation of *interest operators* including a new development, which is based on measuring the distance between sets of *interest points*. In the following section, we will describe the techniques used and their results.

6.3.1 Knowledge Engineering

As we have explained above we use several techniques to capture domain knowledge about the *interest operators*. These techniques are:

- visual inspection;
- ground-truth verification on the basis of good and bad areas;
- evaluation by means of distances between sets of *interest points*.

In the following the several implemented *interest operators* described in Section 6.2 are compared using these three evaluation methods.

6.3.1.1 Visual Inspection

Methods using visual inspection are subjective ones and depend only on the person evaluating the result. We can formulate a set of criteria to evaluate the quality of *interest operators*: number of *interest points*, position of single *interest points* or accumulation of *interest points*. This list may be extended by any other criteria. At the lowest level of inspection human estimates the image and the detected *interest points* and evaluates the ensemble. There are given four image examples (Figure 6.5) to illustrate this.

The *interest operators* in Figure 6.5 were applied with the following input parameters: Förstner 1: $R = 3$; Förstner 2: $q_{min} = 0.2$, $W_{min} = 400$ and $R = 3$; Harris: $\sigma = 1$, $\sigma_I = -0.04$ and $corn_{min} = 0.000005$; HFVM: feature kernel listed in Equation 6.6.

If we compare the four images shown in Figure 6.5, intuitively the best results (points which represent the elementary building facades elements, like edges, windows, beams) are supplied by the Förstner 2 (Figure 6.5b) and the Harris operator (Figure 6.5c). The other two *interest operators* detect many isolated single points. The problem of the Förstner 1 operator (Figure 6.5a) is that the fixed parameters for q_{min} and W_{min} are not suitable for the image at hand. Additionally, the feature set used for the Hierarchical Feature Vector Matching operator is unsuitable.

For the visual inspection, we used about 120 images of building facades. These images are uniformly distributed over the four defined facade-types (see Section 3.4).

The experiences collected in the course of this work can be summarized as follows:

- the Förstner operators are more suitable than the other *interest operators* for non-homogeneous facades;
- the Förstner 1 and the Förstner 2 operator differ only in the required input parameter, but the resulting *interest points* are the same;
- the Harris operator detects points first of all on parts with horizontal or vertical homogeneity;
- the Hierarchical Feature Vector Matching operator only operates satisfactorily with a suitable feature set for each object (facade) type.

6.3 The Knowledge Base – Point Detection

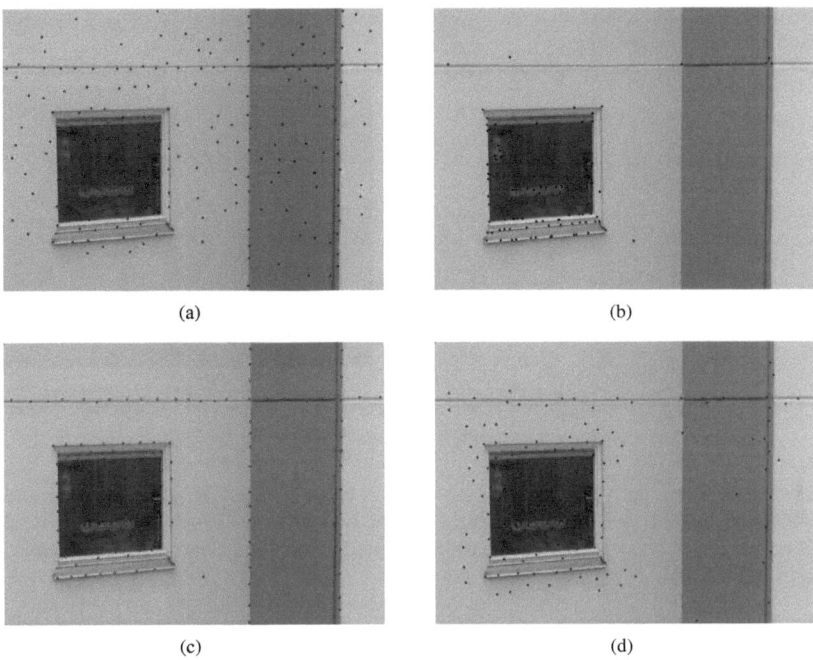

Figure 6.5: Images with extracted points by (a) the Förstner 1 operator; (b) the Förstner 2 operator; (c) the Harris operator and (d) the HFVM operator.

6.3.1.2 Ground-Truth Verification on the Basis of Good and Bad Areas

The ground-truth verification on the basis of good and bad areas determines the correct and false detected *interest points*. This method was developed in cooperation with the *Joanneum Research* in Graz (Austria) and includes the following procedures:

- the user defines several types of areas (masks) in the image;
- application of *interest operators*;
- analysis of the detected points in the context with the defined masks.

Two types of masks have to be defined: the *good-mask* includes the area within which *interest points* are desired; the *bad-mask* includes the area within which *interest points* are not desired.

Based on the defined masks different metrics can be calculated: number of total *interest points* (num_{total}), number of points within the good areas (num_{good}), number of points within the bad areas (num_{bad}) and the actual number to total number ratios:

$$ratio_{good} = \frac{num_{good}}{num_{total}}, \quad ratio_{bad} = \frac{num_{bad}}{num_{total}}. \tag{6.7}$$

This method is also subjective as it uses visual inspection. The masks are produced by humans and the result depends only on them.

(a)   (b)

Figure 6.6: (a) Good-mask and (b) bad-mask for the image from Figure 6.5.

The advantage is that the verification of the result is easier and can be done automatically (if the masks are available). It is noticed that apart from the two areas described above (good-mask / bad-mask), not classified areas (lying outside the mentioned masks) can exist and their different metrics can be calculated, too.

	Förstner 1	Förstner 2	Harris	HFVM
num_{total}	165	102	112	100
num_{good}	4	5	4	3
num_{bad}	161	97	108	97
$ratio_{good}$	0.024	0.049	0.036	0.030
$ratio_{bad}$	0.976	0.951	0.964	0.970

Table 6.1: Verification results for Figure 6.5 using the masks shown in Figure 6.6.

Figure 6.6 shows a good- and a bad-mask for Figure 6.5. The verification results for Förstner 1, Förstner 2, Harris and HFVM are listed in Table 6.1. These values confirm the visual inspection done in Section 6.3.1.1.

6.3.1.3 Evaluation by Means of Distances Between Sets of Points

The two methods described above are based on the validation of the whole scene and on the calculation of *interest points*. Information about localization accuracy is not considered.

The disadvantage of the two methods described above is removed with the evaluation by means of distances between sets of points. This technique was specially developed for the work at hand and builds on calculation methods for point sets used in computer science (Eiter et al., 1997; Ramon, 2001).

The fundamental idea of our evaluation technique is to define two types of point sets:

6.3 The Knowledge Base – Point Detection

- desired points detected by the user (point set 0);
- *interest points* detected by the interest operator to be evaluated (point set 1).

The distance between these two sets of points represents a measure for the quality of the *interest operators* used. First of all we have to define a suitable measure for point sets.

Eiter et al. (1997) proposed the following distance measure (*link distance* d_l) between two sets of points ($S_1, S_2 \subseteq B$):

A so-called *linking* between S_1 and S_2 is a relation $R \subseteq S_1 \times S_2$ satisfying

(a) for every $e_1 \in S_1$ there exists some $e_2 \in S_2$ such that $(e_1, e_2) \in R$, and
(b) for every $e_2 \in S_2$ there exists some $e_1 \in S_1$ such that $(e_1, e_2) \in R$.

The *minimum link distance* d_l between subset S_1 and S_2 is defined by

$$d_l(S_1, S_2) = \min_R \sum_{(e_1, e_2) \in R} \Delta(e_1, e_2), \tag{6.8}$$

where the minimum is taken over all relations R such that R is a linking between S_1 and S_2.

More details and the complete mathematical model can be found in (Eiter et al., 1997; Ramon, 2001). For a better understanding of the *link distance* we will give a simple example (Figure 6.7).

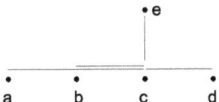

Figure 6.7: Example for the *link distance* measure (symbolic linking).

In Figure 6.7 two sets of points are shown: $S_1 = \{a, b, c, e\}$ and $S_2 = \{c, d\}$. The *link distance* $L = 5$ with $d_l(S_1, S_2) = \{(a, c), (b, c), (e, c), (c, d)\}$.

For the calculation of distances between the single points we use the *Euclidean distance*. In a plane with P_1 at (x_1, y_1) and P_2 at (x_2, y_2), it is:

$$d(P_1, P_2) = \sqrt{(x_1 - x_2)^2 + (y_1 - y_2)^2} \; ; \tag{6.9}$$

where x_1, x_2, y_1, y_2 are the point co-ordinates.

As we have explained the two point sets used for evaluation of an interest operator are the desired points detected by the user (S_1) and the points detected by the interest operator (S_2). The ideal case occurs if $S_1 = S_2$ and therefore the *link distance* $L = 0$. The greater the value for L, the worse the quality of the interest operator.

The problem and disadvantage of this type of evaluation is that an interest operator which correctly detects all desired points plus a few of single isolated points has a greater value of L than an interest operator which extracts only single isolated points near the desired points. This effect is shown in Figure 6.8.

The two point sets for the left image (Figure 6.8a) are:
$S_1 = \{1, 2, 3, 4, 5\}$
$S_2 = \{1', 2', 3', 4', 5', 6', 7'\}$;
for the right image (Figure 6.8b):
$S_1 = \{1, 2, 3, 4, 5\}$
$S_2 = \{1', 2', 3', 4', 5'\}$.

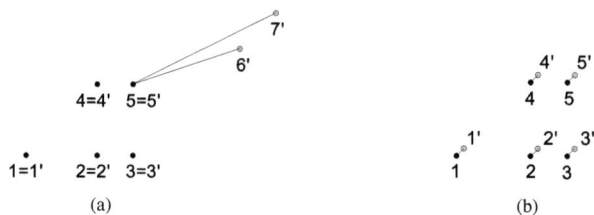

Figure 6.8: A simple example to show the disadvantage of the described evaluation.

The linking for Figure 6.8a is $d_l(S_1, S_2) = \{(1, 1'), (2, 2'), (3, 3'), (4, 4'), (5, 5'), (5, 6'), (5, 7')\}$, which for Figure 6.8b $d_l(S_1, S_2) = \{(1, 1'), (2, 2'), (3, 3'), (4, 4'), (5, 5')\}$.

Provided that the distance from "2" to "3" is the unit distance, the calculated *link distances* are: $L = 0.95$ for the left and $L = 0.36$ for the right image.

The interest operator, which has extracted the points in Figure 6.8a, will be judged worse than the other one in Figure 6.8b. As a matter of fact, this conclusion is intuitively wrong. For this reason we propose an extended measure for evaluation of *interest operators* based on the *link distance*. In the following we call it L_{IOP}.

The fundamental idea is to introduce two "sub-measures" and combine these to a new measure, which yield better results. The "sub-measures" are:

- the *link distance* calculated between the desired points detected by the user (S_1) and all the points detected by the interest operator (S_2): we call it L_1;
- the *link distance* calculated between the desired points detected by the user (S_1) and its nearest neighbor-point: we call it L_2.

The "sub-measures" L_1 and L_2 will be weighted and added:

$$L_{IOP} = w_1 \cdot L_1 + w_2 \cdot L_2 . \tag{6.10}$$

The values for w_1 and w_2 have been specified by means of extensive experiments (calculation of L_1 / L_2 and comparison with the evaluation results obtained from the methods described in Section 6.3.1.1 and 6.3.1.2). For our object types (facades) the following values have been fixed: $w_1 = \frac{1}{5}$ and $w_2 = \frac{1}{2}$.

For the example illustrated in Figure 6.8 the following calculation arises:

Figure 6.8a: $L_1 = 0.95$;
$L_2 = 0$;
thus $L_{IOP} = 0.19$;

Figure 6.8b: $L_1 = 0.36$;
$L_2 = 0.36$;
thus $L_{IOP} = 0.25$.

6.3 The Knowledge Base – Point Detection

In the following we evaluate the example from Figure 6.5. First of all we have to define the desired points (shown in Figure 6.9).

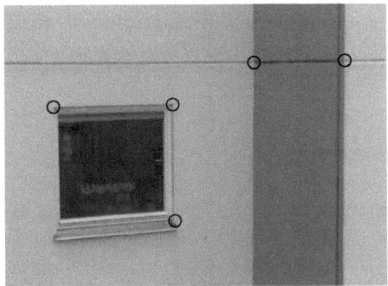

Figure 6.9: Desired points for Figure 6.5.

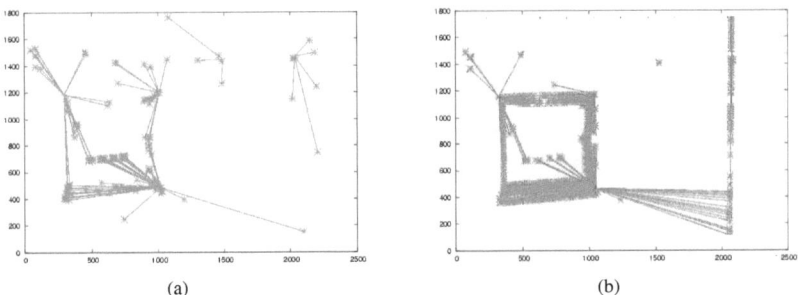

Figure 6.10: (a) The extracted points by the Förstner 2 operator and (b) the Harris operator (including the linking).

After having extracted the points with several *interest operators*, the linking (Figure 6.10) and the calculation of the *link distance* must be done. The results are listed bellow:

Förstner 1 (Figure 6.5a): $L_1 = 34.40$;
$L_2 = 0.92$;
$L_{IOP} = 7.34$;

Förstner 2 (Figure 6.5b): $L_1 = 12.98$;
$L_2 = 2.86$;
$L_{IOP} = 4.03$;

Harris (Figure 6.5c): $L_1 = 19.21$;
$L_2 = 1.32$;
$L_{IOP} = 4.50$;

HFVM (Figure 6.5d): $L_1 = 18.94$;
$L_2 = 2.86$;
$L_{IOP} = 5.22$.

Evaluation on the basis of these values shows that the most convenient *interest operators* for this type of image (facade) are the Förstner 2 and the Harris operator.

The developed evaluation method was tested on about 120 images. The results have confirmed the visual inspection described in Section 6.3.1.1 and the ground-truth verification on the basis of good and bad areas discussed in Section 6.3.1.2. Furthermore these tests have shown that the new "interest operator evaluation method" works sufficiently good.

It is notable that – additionally to the used measure for point sets – there exists a huge number of other measures (Eiter et al., 1997), e.g. Hausdorff distance, surjection distance, fair surjection distance and others. The use of one of these measures instead of the *link distance* yields similar results; the defined values for w_1 and w_2 have to be adapted.

6.3.1.4 Evaluation of the Runtime

Additionally to the evaluation made above we have inspected the runtimes of the interest operators used. This information can be used to choose a suitable operator if more than one is qualified for the situation. The runtimes listed in Table 6.2 are mean values from more then 100 images. The runtime correlates with the detected points and with image size. For our analysis, we have used images with 640×480, 800×600 and 1024×768 pixels (all with 8 BPP). The computer system was a conventional Personal Computer running under Windows 2000 (Intel Pentium 4 - 1.3GHz; 512MB Ram).

	Förstner 1	Förstner 2	Harris	HFVM
640×640	~ 5 sec.	~ 12 sec.	~ 1 sec.	~ 1 sec.
800×600	~ 7 sec.	~ 23 sec.	~ 1 sec.	~ 1 sec.
1024×768	~ 15 sec.	~ 45 sec.	~ 3 sec.	~ 2 sec.

Table 6.2: Results of runtime analysis (P4 - 1.3GHz; 512MB Ram).

6.3.1.5 Evaluation-Results

As mentioned above we used about 120 images of building facades for the evaluation. These images are uniformly distributed over the four defined facade-types (see Section 3.4). The evaluation process was done with the three methods described including the runtime analysis.

Cluster analysis was also used to determine coherences between image features and suitable *interest operators*. *Cluster analysis* is a multivariate analysis technique that seeks to organize information about variables so that relatively homogeneous groups or "clusters" can be formed. We have used the so-called *Ward's method*. This method is distinct from all other methods because it uses an analysis of variance approach to evaluate the distances between clusters. We refer to Bortz (1999) for details concerning this method. In general, this method is regarded as very efficient, however, it tends to create clusters of small size. The results of various *cluster analyses* done confirm the results of the other evaluation methods. First of all, the image features used for the decision process were reconfirmed.

The results of all evaluation techniques can be summarized as follows:

- the Förstner 2 operator is more suitable than the other *interest operators* for non-homogeneous facades;
- the Förstner 1 and the Förstner 2 operator differed in the required input parameter, but the resulting *interest points* are the same;

6.3 The Knowledge Base – Point Detection

- due to more input parameters, the Förstner 2 can be adapted flexibly to the current situation;
- the Förstner 1 has a shorter runtime than Förstner 2;
- the Harris operator is suitable for images (or image parts) with horizontal or vertical homogeneity;
- the Hierarchical Feature Vector Matching (HFVM) operator only operates satisfactorily with a suitable feature set;
- the used feature set (see Equation 6.6) for the HFVM operator is only suitable for steel-glass facades;
- the HFVM operator is unsuitable if reflections or/and shadow cast is on the facade;
- the parameters $corn_{min}$ for the Harris and W_{min} for the Förstner 2 operator have to be chosen depending on the contrast of the image.

These collected evaluation results are the basis for the formulated rules described in the next sections. As we have explained at the beginning of this section, this rule base is divided into three groups of rules: rules for the choice of a suitable interest operator, rules for the predefinition of necessary parameters and rules to define the order of the algorithms.

We will list the preconditions in form of tables in order not to anticipate the concrete implementation of the rules and their syntax. A listing of the complete knowledge base can be found in Appendix C.2. The syntax of a rule and its relevant concept has been described in Section 4.2.1.

6.3.2 Rules for the Choice of Suitable *Interest Operators*

(a) the *Förstner 1 interest operator* is suited for non-homogeneous facades as described above. This can be checked by means of the first and fifth Haralick feature. If these two values are small and/or the third histogram feature (*skewness*) has a very high negative value, the Förstner 1 operator can be applied. The negative *skewness* value indicates an image with stronger bright areas. The necessary preconditions for the Förstner 1 interest operator are listed in Table 6.3.

(b) the *Förstner 2 interest operator*, as the Förstner 1, is suited for non-homogeneous facades. A difference between the two operators is that the Förstner 2 interest operator will not be applied if the *skewness* has a negative value. If both operators are suited, the rules, that will be described in Section 6.3.4, define the order of the algorithms. The necessary preconditions for the Förstner 2 interest operator are listed in Table 6.3.

Förstner 1 interest operator	
$H_1 - 0°$	very low *or* low *or* middle
$H_5 - 0°$	very low *or* low *or* middle
or	
$H_1 - 90°$	very low *or* low *or* middle
$H_5 - 90°$	very low *or* low *or* middle
or	
$H_1 - 0°$	very low *or* low
$H_5 - 90°$	very low *or* low
M_3	very high negative
or	
$H_1 - 0°$	very low *or* low
$H_5 - 90°$	very low *or* low
M_3	very high negative

Förstner 2 interest operator	
$H_1 - 0°$	very low *or* low *or* middle
$H_5 - 0°$	very low *or* low *or* middle
or	
$H_1 - 90°$	very low *or* low *or* middle
$H_5 - 90°$	very low *or* low *or* middle
or	
$H_1 - 0°$	very low *or* low
$H_5 - 90°$	very low *or* low
M_3	not very high negative
or	
$H_1 - 0°$	very low *or* low
$H_5 - 90°$	very low *or* low
M_3	not very high negative

Table 6.3: Preconditions for the Förstner 1 and Förstner 2 interest operator.

(c) the *Harris interest operator* is suitable for images with horizontal or vertical homogeneity. This can be checked by means of the first and fifth Haralick feature. If these two values for one direction are relatively high, the Harris interest operator can be applied. The necessary preconditions for the Harris operator are listed in Table 6.4.

(d) the *Hierarchical Feature Vector Matching (HFVM) operator* with the developed feature set is only suitable for steel-glass facades. Additionally no reflections or/and shadow cast are permitted. These image features are asked in form of user-queries (UQ) described in Section 3.4. The necessary preconditions for the HFVM operator are listed in Table 6.4.

6.3.3 Rules for the Predefinition of Necessary Parameters

(a) the *Förstner 1 interest operator* works with fixed parameters for q_{min} and W_{min}. Only the window length R can be chosen by the user, respectively by the system. In the current imple-

6.3 The Knowledge Base – Point Detection

Harris interest operator	
$H_1 - 0°$	very high *or* high *or* middle
$H_5 - 0°$	very high *or* high *or* middle
or	
$H_1 - 90°$	very high *or* high *or* middle
$H_5 - 90°$	very high *or* high *or* middle

HFVM operator	
UQ1 - type of facade	steel-glass
UQ2 - reflections on the facade	none
UQ3 - shadow cast on the facade	none

Table 6.4: Preconditions for the Harris and the HFVM interest operator.

mentation we have fixed $R = 3$.

(b) the *Förstner 2 interest operator* permits the selection of values for q_{min}, W_{min} and R. Since we use this interest operator for finding corner points we have fixed $q_{min} = 0.2$ and $R = 3$. W_{min} will be chosen depending on the contrast (Haralick features - H_2 (average) and first histogram feature *mean* - M_1) of the image. The coherences are listed in Table 6.5.

M_1	H_2 (avg)	W_{min}
very low	very low	32
very low	low	36
very low	middle	40
very low	high	45
very low	very high	50
low	very low	176
low	low	198
low	middle	220
low	high	248
low	very high	275
middle	very low	320
middle	low	360
middle	middle	400
middle	high	450
middle	very high	500

M_1	H_2 (avg)	W_{min}
high	very low	480
high	low	540
high	middle	600
high	high	675
high	very high	750
very high	very low	640
very high	low	720
very high	middle	800
very high	high	900
very high	very high	1000

Table 6.5: Parameter W_{min} for the Förstner 2 interest operator.

(c) the *Harris interest operator* permits the selection of values for σ, N_0 and $corn_{min}$. Since this interest operator is suited for images with horizontal or vertical homogeneity we have fixed $\sigma = 1.0$ and $N_0 = -0.04$. $corn_{min}$ will be chosen depending on the contrast (Haralick features - H_2 (average) and first histogram feature *mean* - M_1) of the image. The coherences are listed in Table 6.6.

(d) the *Hierarchical Feature Vector Matching (HFVM) operator* needs a suitable feature set as input. We use only one feature (see Equation 6.6), therefore in our implementation the HFVM operator has no parameter.

M_1	H_2 (avg)	$corn_{min}$
very low	very low	0.00005
very low	low	0.0001
very low	middle	0.00015
very low	high	0.00023
very low	very high	0.0003
low	very low	0.00006
low	low	0.00095
low	middle	0.0018
low	high	0.0028
low	very high	0.0037
middle	very low	0.00007
middle	low	0.0018
middle	middle	0.0035
middle	high	0.0053
middle	very high	0.007
high	very low	0.000085
high	low	0.0022
high	middle	0.00425
high	high	0.0064
high	very high	0.0085
very high	very low	0.0001
very high	low	0.0026
very high	middle	0.005
very high	high	0.0075
very high	very high	0.01

Table 6.6: Parameter $corn_{min}$ for the Harris interest operator.

6.3.4 Rules to Define the Order of the Algorithms

These rules go through the list of chosen *interest operators* and decide which actions have to be carried out in which order. The knowledge generally results from the runtime evaluation (see Section 6.3.1.4) and experiences.

(a) *Förstner 1 interest operator* is always the first even if additionally the Förstner 2 interest operator is suited;

(b) *Förstner 2 interest operator* is the first operator or the first after the Förstner 1;

(c) *Harris interest operator* is the first operator after the Förstner 1 / Förstner 2 operator;

(d) *Hierarchical Feature Vector Matching (HFVM) operator* is always the last operator to be applied.

6.3.5 Overview – Rulebase

Due to the fact that only four *interest operators* were implemented, the knowledge base could be kept relatively simple and thus easily modifiable and extendable.

In the following we will give some additional information about the knowledge base developed.

The complete knowledge base for the choice of a suitable interest operator consists of 19 rules.

The three "sub rule bases" are:

- rules for the choice of suitable algorithms – 5 rules;
- rules for the predefinition of necessary parameters – 4 rules;

6.4 Examples

- rules to define the order of the algorithms – 4 rules.

As the knowledge base for the choice of suitable image preprocessing and image enhancement algorithms, this knowledge base also contains rules to display and write the necessary output and rules to document conclusions (a total of 6 rules). A listing of the complete knowledge base can be found in Appendix C.2.

6.4 Examples

In this section we will, by means of two examples, provide a better understanding of the functionality of the system developed. We assume that the quality of the images is suitable for the detection of *interest points*. Such a quality results from a high qualitatively captured image or from performed image preprocessing and image enhancement. As mentioned at the beginning of this chapter building facades elements can be represented by simple line geometry. Therefore the process of point detection can be reduced on capturing points along such lines and intersections of them.

6.4.1 First Example:

This first example shows a steel-glass facade with strong reflections on the right side (Figure 6.11). First of all image analysis has to be done. The extracted image features are listed in Table 6.7. The user questions (UQ) (see Section 3.4) have been answered as follows:

- UQ1: What kind of type is the facade? → steel-glass facade (= type D).
- UQ2: Are there any reflections on the facade? → very strong.
- UQ3: Is there a shadow cast on the facade? → none.

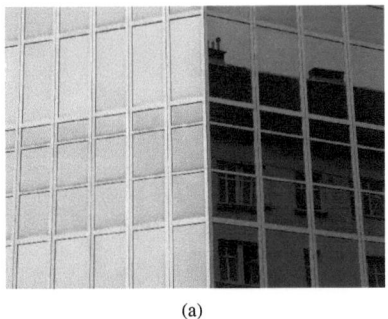
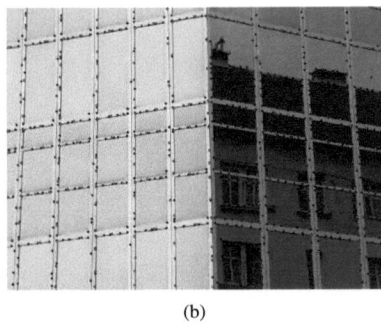

(a)   (b)

Figure 6.11: (a) Image with strong reflections on the right side; (b) *interest points* detected with the Förstner 2 and the Harris operator.

If we compare the preconditions of the rules described (Section 6.2) with the extracted image features from Table 6.7, the following interest operators are selected: Förstner 2 and the Harris operator.

M_1	M_1 - f.v.	M_2	M_2 - f.v.	M_3	M_3 - f.v.				
135.7255	middle	75.5022	low	-0.33192	low negative				
	0°	f.v.	45°	f.v.	90°	f.v.	135°	f.v.	average	f.v.
H_1	0.002	mid.	0.002	mid.	0.003	high	0.002	mid.	0.002	mid.
H_2	397.052	v.high	600.586	v.high	236.372	high	621.216	v.high	463.807	v.high
H_5	0.450	high	0.389	mid.	0.463	high	0.389	mid.	0.423	high
H_9	3.165	mid.	3.291	mid.	3.080	mid.	3.287	mid.	3.206	mid.

Table 6.7: Extracted image features for Figure 6.11a.

Figure 6.11b shows the applied *interest operators*: Förstner 2 with $q_{min} = 0.2$, $W_{min} = 400$ and $R = 3$; Harris with $\sigma = 1.0$, $N_0 = -0.04$ and $corn_{min} = 0.0035$. The parameters have been chosen on the basis of the rules defined in Section 6.3.3.

It can be seen that points are generally detected on the steel structure of the facade and only a small number of isolated single points are detected on the glass windows. These points result from local grey-level differences, like "fault-pixels". An exception to this is the right part of the image, where many *interest points* are caused by the reflections on the facade. Also a variance of the chosen parameters can not eliminate this problem, since the grey-level differences in this area are the same as those of the steel structure of the facade. Therefore changes of parameter values remove the undesirable points on the glass windows, but also the desired *interest points*. Undesirable points can only be eliminated by a suitable point filtering technique (see Section 6.5.2).

6.4.2 Second Example:

This second example (Figure 6.12) shows, similar to the example shown in Figure 6.11, a steel-glass facade, but the structure of the building is more regular than the other one and fewer reflections are present.

(a) (b)

Figure 6.12: (a) Steel-glass facade; (b) *interest points* detected with the HFVM and the Harris operator.

As we have explained above first of all image analysis must be done. The extracted image features are listed in Table 6.8.

6.4 Examples

M_1	M_1 - f.v.	M_2	M_2 - f.v.	M_3	M_3 - f.v.				
186.4555	v. high	21.3357	v. low	-0.3539	low negative				
	0°	f.v.	45°	f.v.	90°	f.v.	135°	f.v.	average	f.v.
H_1	0.006	v.high	0.005	high	0.006	v.high	0.005	high	0.005	v. high
H_2	90.105	low	134.469	mid.	40.936	v.low	134.070	mid.	99.895	low
H_5	0.527	high	0.419	high	0.514	high	0.423	high	0.471	high
H_9	2.757	low	2.955	low	2.738	low	2.947	low	2.849	low

Table 6.8: Extracted image features for Figure 6.12a.

The user questions (UQ) (see Section 3.4) have been answered as follows:

- UQ1: What kind of type is the facade? → steel-glass facade (= type D).
- UQ2: Are there any reflections on the facade? → none.
- UQ3: Is there a shadow cast on the facade? → none.

On the basis of the extracted image features from Table 6.8 the following *interest operators* are selected by the rules (Section 6.2): HFVM and the Harris operator.

Figure 6.12b shows the applied *interest operators*: Harris with $\sigma = 1.0$, $N_0 = -0.04$ and $corn_{min} = 0.0026$, and the HFVM with the feature set defined in Equation 6.6. As, just described in *Example 1*, it can be seen that points are generally detected on the steel structure of the facade. Some of the shutters behind windows result in a small number of *interest points* on the glass areas. These points are caused by the same effect as the reflections in *Example 1*; a variance of the chosen parameters can not remove these single points, since the grey-level differences in this area are the same as those of the steel structure.

In this example the choice of suitable *interest operators* generally depends on the user questions. This can be shown by modifying the answers as follows:

- UQ1: What kind of type is the facade? → steel-glass facade (= type D).
- UQ2: Are there any reflections on the facade? → middle.
- UQ3: Is there a shadow cast on the facade? → none.

The second user question has been changed from "none" to "middle". Due to the modified image features instead of the HFVM operator the Förstner 2 operator ($q_{min} = 0.2$, $W_{min} = 1000$, $R = 3$) has been chosen. The resulting *interest points* are shown in Figure 6.13.

Comparing Figure 6.12 and Figure 6.13 it can be seen that the results are nearly the same. Some diversities are:

- the combination Förstner 2 / Harris operator detects more *interest points* than the HFVM / Harris;
- Förstner 2 / Harris detect some points twice (the Förstner 2 as well as the Harris operator);
- HFVM / Harris detect more isolated single points on window-areas than the Förstner 2 / Harris combination.

Figure 6.13: *Interest points* detected with the Förstner 2 and the Harris operator.

In conclusion it can be said that the results from Förstner 2 / Harris operator, shown in Figure 6.13, are slightly better than the results from HFVM / Harris operator. This confirms the necessity of user questions in addition to the extracted image features.

Furthermore can be said that in spite of choosing suitable algorithms for image preprocessing and image enhancement and for *interest operators*, the number of points is mostly very high. Therefore the next sequence steps have to be the reduction of the interest points by a special point filter. This will be described in the next section.

6.5 Knowledge-Based Point Filtering

As we have described above, the knowledge-based system chooses a suitable *interest operator* on the basis of the extracted image features. In normal cases not only one *interest operator* will be selected, but a group of suitable algorithms. Therefore in the course of finding interest points, more than one *interest operator* will be applied. This results in single lists of *interest points*.

The point filter, which will be described in the following, has the task to unite the single lists and to reduce the points according to certain rules.

Point reduction is necessary since, in spite of choosing suitable algorithms for image preprocessing and image enhancement and for *interest operators*, many undesirable points (see the examples in Section 6.4) are detected. The application of the developed measurement system is focused on monitoring of building facades. The point detection has to be done in such a way, that the extracted points characterize the object. In case of facades the elementary structure can be represented by a simple line geometry. Points detected apart from this line structure (e.g. points inside glass windows) are undesirable and not useful for subsequent process steps, like object reconstruction or deformation analysis.

Thereto we expand the architecture of our system shown in Figure 6.1 by the knowledge-based point filtering (Figure 6.14), which has as input the single point lists obtained from applied *interest operators*.

The point reduction is not a removing process of undesirable points (all detected points will be preserved, in avoidance of loss of information), but is based on the weighting of each point.

6.5 Knowledge-Based Point Filtering

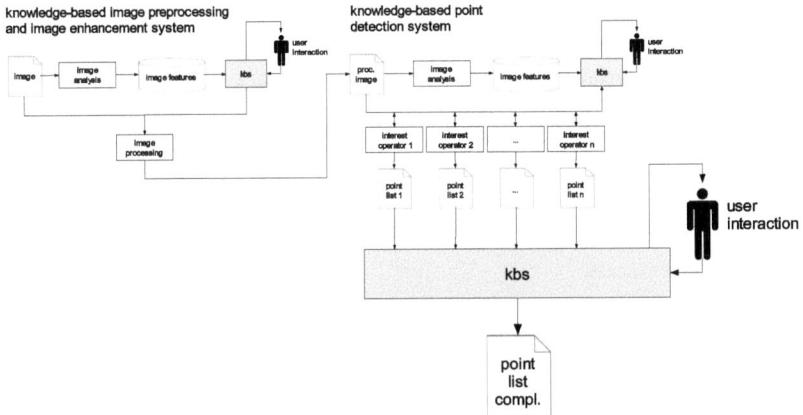

Figure 6.14: Architecture of knowledge-based point filter (including the preceding processing steps).

The filtering process will be done by means of two methods:

- point filtering on basis of defined rules,
- interactive point filtering.

In the following, we will describe these two methods in more detail.

6.5.1 Point Filtering on the Basis of Defined Rules

In this first method several criterions will be used for weighting each point, which can be summarized as follows:

(a) how many *interest operators* detect the point and

(b) which "property-parameters", obtained from the corresponding *interest operator*, the point has.

The *first criterion* is very simple but effective. The point filter scans all point lists (one point list for each applied *interest operator*) and weights each point in correspondence with the number of *interest operators*, from which this point has been detected. In practice this is a search routine which finds points with the same co-ordinates in different point lists. The weights are fixed on the basis of this simple coherence. More details how the weights are determinated can be found in Section 6.5.3.

The *second criterion* is based on "property-parameters" (available in our implementation) obtained from the corresponding *interest operator* for each point:

- the Förstner 1 and Harris operator return the standardized grey value (between 0 and 1) for the correspondent point;
- the Förstner 2 operator provides the values for q and W (see Equation 6.1) for each point ;

- the HFVM operator provides the value of the sum of grey-value-variances to neighbors of a 7×7 mask for each point; the definite value arises from the standardized sum (between 0 and 1) from the first to the highest level.

On the basis of these returned values we can formulate several rules for point filtering. In the simplest case threshold values will be used. Points with returned values less than the threshold values get a different weight from points with returned values greater than or equal to the thresholds. Following points with low weight can be removed.

This technique is realised in our system. In Table 6.9 relevant threshold values are proposed. They are obtained from experiments:

Förstner 1	Förstner 2	Harris	HFVM
$gv_{th} = 0.01$	$W_{th} = 1000$, $q_{th} = unused$	$gv_{th} = 0.1$	$gv_{th} = 0.1$

Table 6.9: Threshold values for point filtering.

For a better understanding of the functionality, we will provide an example:

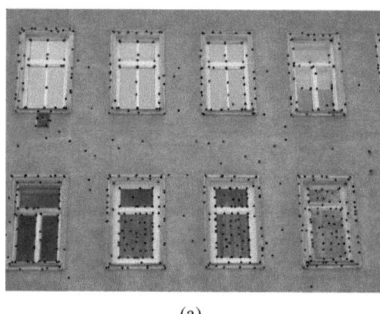   

(a)   (b)

Figure 6.15: (a) Extracted points by the Förstner 2 operator and (b) the detected points after filtering.

In Figure 6.15a points detected by a single interest operator, the Förstner 2 operator, are shown. It can be seen that the result is of average quality. A few of the *interest points* are suitable for subsequent analysis (e.g. deformation analysis, reconstruction, and others), but many of them represent isolated points on the masonry or on the window-areas. This effect results from high local grey value variations present in this area, which could not be represented and detected with the developed image analysis (see Chapter 3).

A intuitively better result (points which represent the elementary building facades elements, like edges, windows, beams) can be obtained after point filtering. In this example we have detected points with the Förstner 2 *interest operator*. Therefore the filter removes points with $W_{actual\ point} < 1000$. However, it should be noted that the removal process is not an erasing, but only a weighting one.

The resulting image is shown in Figure 6.15b. Most of the isolated points on the masonry have been filtered, even though some undesirable *interest points* remained (e.g. the window-areas on the bottom right).

To eliminate such undesirable point clouds, we have developed an interactive point filter described in the next section.

6.5.2 Interactive Point Filtering

The developed interactive point filter allows the user to choose the points or point clouds to be removed. This selection process is realised by means of a graphical user interaction, shown in Figure 6.16a.

The user has to draw a rectangular window in the graphical output. Points inside these selected windows will be removed. More details about the practical implementation can be found in Section 8.

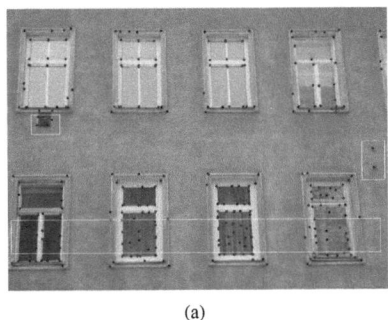
(a)

(b)

Figure 6.16: (a) User–graphic interaction for point filtering; (b) points after interactive point filtering.

In a final step the user has the possibility to let show only points with the same weight and to remove point groups which are weighted differently. For example, it is possible to keep points with weight 2, 4 and 5.

This method enables the user to adjust the *interest points* to a best qualification for all sorts of subsequent applications.

6.5.3 Weighting of Filtered Points

As we have explained in the previous section the developed filtering technique is based on weighting of selected points. This selection can be done automatically by defined rules or interactively by the user.

The weight will be determined as follows:

- **step 1:** point detected by one *interest operator* $\rightarrow weight = 1$, point detected by two *interest operators* $\rightarrow weight = 2$, and so on;
- **step 2:** points with returned values less than the threshold values defined in Table 6.9 get the same weight as before, points with returned values greater than or equal to the thresholds get a new weight: $weight_{new} = weight_{old} + 1$;
- **step 3:** points selected by the user to be removed get the new $weight = -1$;

It should be noted that the developed point filtering technique is a very simple method and only suitable for building facade images. An "intelligent point filtering" has lot of potential for future extensions of the system.

6.5.4 Examples

In this section, we will provide a better understanding of the functionality of the developed point filter by means of two examples. For each example we will list four images: original image, detected points without filtering, detected points after the rule-based filtering and detected points after the complete filtering process.

6.5.4.1 First Example:

This first example shows the image from Figure 6.11b. The points have been detected with the Förstner 2 and the Harris operator (more details about the chosen *interest operators* and their parameters can be found in Section 6.4.1). The reflections on the facade cause many *interest points* on the right side part of the glass structure (Figure 6.17b).

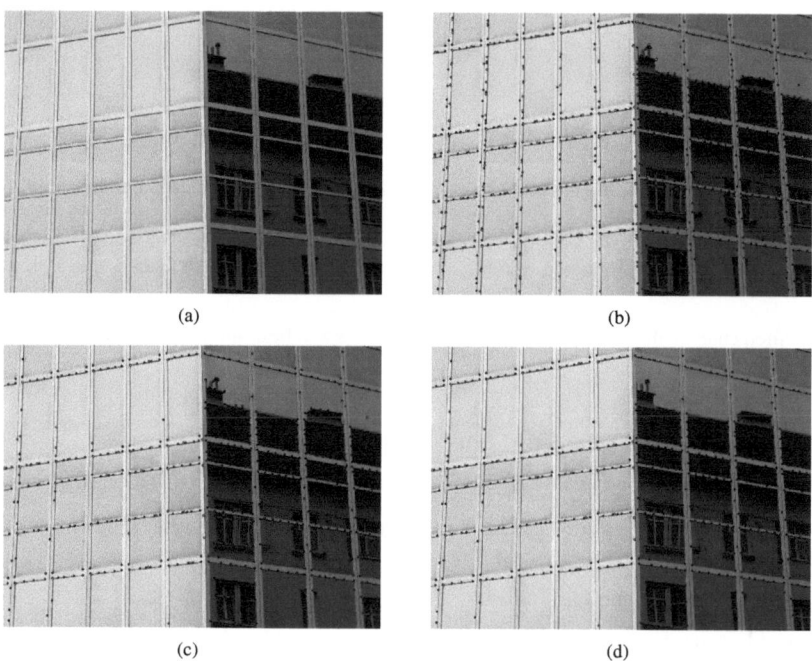

Figure 6.17: (a) Image with strong reflections on the right side; (b) *interest points* detected with the Förstner 2 and the Harris operator; (c) points after rule-based point filtering; (d) points after interactive point filtering.

In this example no points have been detected by more than one *interest operator*. Therefore, in the first step the weights remain unchanged ($weight = 1$). In the second step the returned values by the *interest operators* are inspected. Points which have been detected by the Förstner 2 operator and which have a returned value less than $W_{th} = 1000$ (see Table 6.9), get the same weight as

6.5 Knowledge-Based Point Filtering

before ($weight = 1$). Points with returned value greater than or equal to this threshold receive a new weight: $weight_{new} = weight_{old} + 1 = 2$. Points which have been detected by the Harris operator are weighted by means of the threshold $gv_{th} = 0.1$. Figure 6.17c shows the points weighted with 2. After this filtering the number of isolated points on the glass areas is reduced. In the third step the user has the possibility to select other points to be filtered and to adapt the point cloud ideally for subsequent process steps (e.g. object reconstruction or deformation analysis). A possible resulting image is shown in Figure 6.17d. Undesirable points inside the glass windows on the right side of the facade have been filtered by the user.

6.5.4.2 Second Example:

This second example (Figure 6.18) shows a steel-glass facade including *interest points* detected by the Förstner 1 and the HFVM operator (Figure6.18b). The point cloud contains many undesirable *interest points*, particularly inside the glass areas.

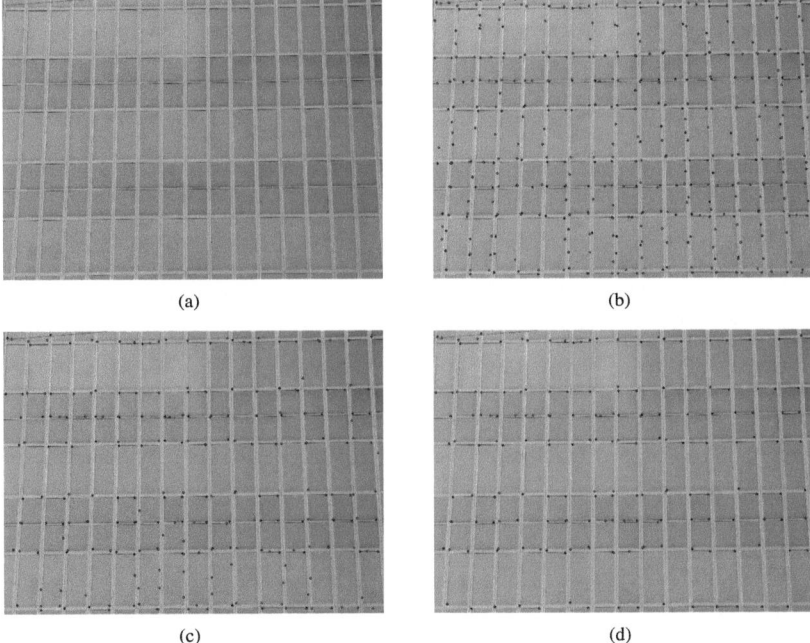

Figure 6.18: (a) Steel-glass facade; (b) *interest points* detected with the Förstner 1 and the HFVM operator; (c) points after rule-based point filtering; (d) points after interactive point filtering.

Just as in *Example 1* there exist no points which have been detected by more than one *interest operator*. Therefore in the first step the weights remain unchanged ($weight = 1$), too.

In the second step, points which have been detected by the Förstner 1 operator and which have a returned value less than $gv_{th} = 0.01$ (see Table 6.9), get the same weight as before ($weight = 1$).

Points with returned value greater than or equal to this threshold get a new weight: $weight_{new} = weight_{old} + 1 = 2$. Points which have been detected by the HFVM operator are weighted by means of the threshold $gv_{th} = 0.1$. Figure 6.18c shows the points weighted with 2.

After the first and second step (Figure 6.18c), most of the points inside the glass areas have been filtered. Similarly to the previous example the user has the possibility to remove the points not suited for subsequent applications (e.g. points inside the glass areas), by means of the interactive point filter (third step). A possible resulting image is shown in Figure 6.18d.

6.6 Summary

In this chapter we have presented a knowledge-based method for choosing *interest operators*. This process step follows directly after image preprocessing and image enhancement has been done. The choice of suitable algorithms for finding *interest points* is based again on extracted image features (histogram features and Haralick features).

The area of application of the developed measurement system is focused on monitoring of building facades. Therefore the developed knowledge base and the described examples are fitted according to this.

The implementation of different interest operator algorithms results in a flexible measurement system which is able to react to several situations (types of images, light situations, and others).

The application of *interest operators* is a computing intensive process. Therefore the runtimes of these steps are longer than those of the image preprocessing and image enhancement system. Whatever, the system shows a reasonable performance for an on-line measurement system (less than 50 seconds for a picture including image analysis).

Additionally in this chapter we have presented a knowledge-based point filter, which is based on the weighting of detected *interest points*.

The point reduction has to be done in such a way, that the resulting points characterize the object. In our case (building facades) the elementary structure can be represented by a simple line geometry. Points detected apart from this structure are undesirable and not useful for subsequent process steps.

Point selection can be done automatically by defined rules or interactively by the user. This approach results in a flexible and easy extendable point filtering technique. The point filter represents the final step in the process sequence and results in a weighted point list for each image.

The developed knowledge-based point finding system was tested on about 120 images. The resulting (filtered) *interest points* represents the object intuitively better than the unfiltered point cloud and are more suitable for subsequent process steps, like object reconstruction or deformation analysis.

On the basis of the resulting (filtered) *interest points* subsequent processing steps can be carried out, like deformation analysis. Deformation analysis is based on the observation of characteristic points in certain time intervals. A condition for this process is the observation of the same object points (object elements) in different measurement epochs. Some basic approaches to find corresponding points respectively elements will be described in the next chapter.

Chapter 7

Image Matching Strategies

Image matching is one of the central tasks in computer vision. Matching aims establishing the relation between two images. Image matching strategies are applied when features, like points, lines or segments are transferred from one image to an other. In this chapter we give an overview of the core methods for matching, which should not be seen as complete treatise of this matter. More details about image matching can be found in (Förstner, 1995; Paar et al., 1996; Kolesnik et al., 1998; Paar et al., 2001).

7.1 Introduction

To solve the matching problem several methods have been developed and published. The best known and most common techniques use local correlation coefficients to describe local similarities.

We use a matching technique to find corresponding points in two or more measurement epochs. This is necessary if, as a subsequent step, deformation analysis has to be done by means of the captured images. Modeling the deformation of an object means to observe the characteristic points in certain time intervals (epochs) in order to properly monitor the temporal course of the movements. A condition for this process is the observation of the same object points in different measurement epochs.

Generally, the matching algorithms can be divided into three groups of methods (Paar et al., 2001):

- raster-based matching,
- feature-based matching,
- relational matching.

In the following we will discuss these three methods in more detail.

7.2 Raster-Based Matching

Raster-based matching methods use the grey-levels themselves as description of the images. The goal of this matching method is to find a mapping function between the reference image and the search image. This can be done by comparing the grey-levels of the two images. Approaches for the location of corresponding image parts are the cross correlation and the Least Squares Matching (LSM). In the following we used for our approach the cross correlation method. Therefore we will confine to the description of this method. More details about the LSM method can be found in (Paar et al., 2001).

Cross correlation method: The reference image I_R (epoch 0) is moved in the search image I_S (epoch 1), and the cross correlation coefficient $k_{R,S}$ of grey-levels will be calculated as follows (Paar et al., 2001):

$$k_{R,S}(\Delta r, \Delta c) = \frac{\sum_{r_R,c_R} [g_R(r_R, c_R) - \bar{g}_R] \cdot [g_S(r_R + \Delta r, c_R + \Delta c) - \bar{g}_S]}{\sqrt{\sum_{r_R,c_R} [g_R(r_R, c_R) - \bar{g}_R]^2 \cdot \sum_{r_R,c_R} [g_S(r_R + \Delta r, c_R + \Delta c) - \bar{g}_S]^2}}, \qquad (7.1)$$

where g_R and g_S are the grey values for reference and search image, $\bar{g}_R$ and $\bar{g}_S$ their mean grey values, Δr and Δc the shift values (see Figure 7.1).

Paar et al. (2001) have simplified Equation 7.1 as follows:

$$k_{R,S} = \frac{\sum g_R \cdot g_S - \sum g_R \cdot \sum g_S}{\sqrt{[\sum g_R^2 - (\sum g_r)^2] \cdot [\sum g_S^2 - (\sum g_s)^2]}}, \qquad (7.2)$$

where $\sum g_R$ and $[\sum g_R^2 - (\sum g_r)^2]$ are constant.

The principle of cross correlation is shown in Figure 7.1; the reference image (in this example the black small part) is moved in the search image (in this example a part of a building facade). The cross correlation coefficient will be calculated as described above.

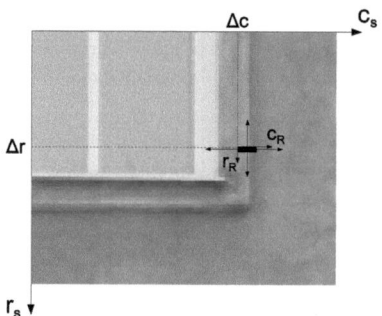

Figure 7.1: Cross correlation.

Subpixel estimation is also possible with cross correlation matching. More details about this approach can be found in (Kolesnik et al., 1998).

We have extended the cross correlation methods by a subsequent step: the places with maximum similarity are combined with the points detected by the *interest operators*. In the neighborhood of maximum correlation, the interest point with the minimum distance to the correlation peak is chosen. This point represents the correspondent point.

It arises a matching process, which consists of the following two steps:

7.2 Raster-Based Matching

- using the cross correlation to describe the local similarities (first step);
- using the *interest points* to specify co-ordinates (second step).

This approach has two essential advantages in comparison with a single correlation matching process:

- we receive subpixel location using the information from the *interest operators*;
- we avoid inaccuracies from subpixel estimation by the matching.

The matching technique described was developed in cooperation with the *Joanneum Research* in Graz (Austria).

The developed two-step matching routine has been tested with a huge number of image pairs of various facades. Each image pair consists of an undeformed image (the first epoch) and a deformed image (the second epoch). The deformed images have been created by moving one or more parts of the original images or by scaling respectively distorting the whole images. These processing steps have been done by means of picture manipulation software (we used Gimp).

To show the effectiveness of the matching algorithm in the following we will present an example (Figure 7.2). Figure 7.2a shows the undeformed original image. By moving downward the upper window row (about 2 pixels) of this image a deformation of the building facade is simulated (Figure 7.2b). First of all point detection (for this example with the Förstner 2 operator) has to be performed, which results in two separated point lists (one for each image / epoch). The goal of the matching routine is to find the correspondent points in these two point lists.

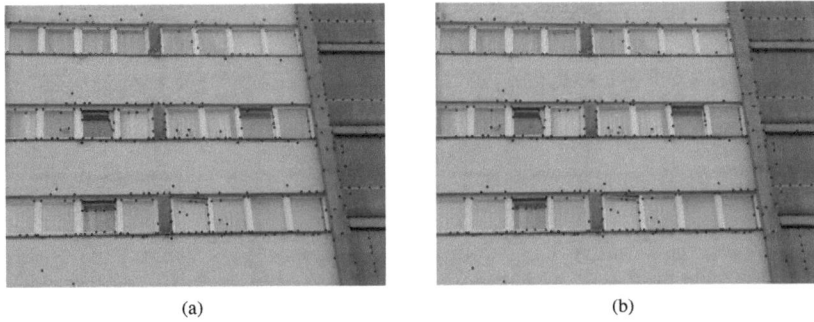

(a) (b)

Figure 7.2: (a) Points detected by the Förstner 2 operator; (b) matched points.

The matching result is shown in Figure 7.2b. A great number of *interest points* has been matched, including points on the deformed part. A condition for the successful matching of a point is that the same object point has been detected in both images. Points which exist only in one image cannot be matched. Examples for such points are the object points upside the moved image part in Figure 7.2a and b.

7.3 Feature-Based Matching

The *feature-based matching methods* do not use the grey-levels as description for the images. These techniques use an abstract image representation form, derived from feature extraction algorithms. The types of used features depend on the object (image) to be matched. Edge elements, corners, line segments, and curve segments are features that are robust against the change of perspective, and they have been widely used in many stereo vision works. Paar et al. (2001) give a (incomplete) list of different feature extraction schemes:

- contour-based primitives (straight line segments, curve segments, corners);
- region-based primitives (region segmentations, homogeneous blobs);
- 3D-surface primitives;

Feature extraction is done by suitable feature extraction algorithms, like edge detectors, corner detectors, and others.

After having extracted the image features, the original raster images are replaced by a symbolic description. The goal of the *feature-based matching process* is to detect homologous features in two (or more) images. The stereo *feature-based matching process* is a very difficult search procedure. In order to minimize false matches, some matching constraints have to be used, e.g. similarity of grey-levels (the matching pixels must have similar grey-level values) or epipolar constraints (given a feature point in the reference image, the corresponding feature point in the search image must lie on the corresponding epipolar line – this constraint reduces the search space from two-dimensional to one-dimensional).

The problem of finding the correspondence between homologous features from different images can be solved by an affine transformation.

We restrict our attention to building facades which are rich in line segments. Therefore the matching complexity can be reduced by matching lines instead of edge pixels, because a collection of pixels can be compared (matching constraint). Figural continuity constraint is automatically enforced because straight edge contours are treated as units.

Details about this matching method can be found in (Paar et al., 2001). The principle of *feature-based matching* is shown in Figure 7.3.

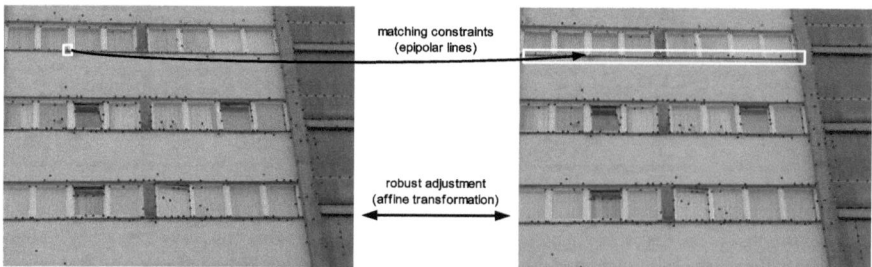

Figure 7.3: The principle of *feature-based matching*.

7.4 Relational Matching

Relational matching is the most powerful matching technique available, as it refers to a symbolic description of the image. High level image analysis deals with image features, like points, lines or surface patches, which are more abstract than image pixels. Such features in a scene are not only defined by properties about the features themselves, but also related to each other by relations between them. Therefore, an object or a scene is represented by features constrained by the properties and relations.

We will discuss the relevant concept of *relational matching* by means of a (simple) example. We take a simplified facade as shown in Figure 7.4a, which has been segmented into different regions (e.g. a and d are concrete beams; b is a masonry; c and e are windows).

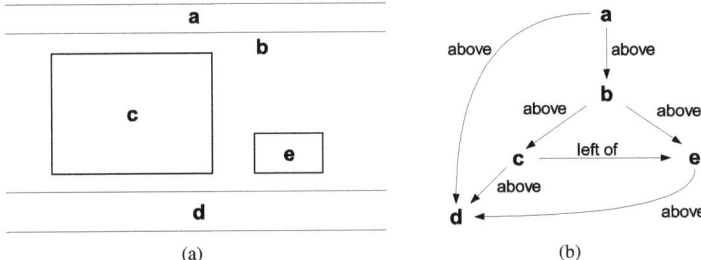

Figure 7.4: (a) Segmented image of a facade with named regions as features; (b) the relations between the segmented regions.

The description D of the image consists of a list of attributed primitives, like color or form: $D = (color, form, ...)$. The relations between the regions can be realized as tree (Figure 7.4b), the *relational matching* as tree search method. Possible relations are: above, left of, right of, in front of, and others. The most general way to find a mapping is to completely scan the tree in a systemic manner called blind search, namely Depth-First and Breadth-First search. It is notable that usually the search time is long and very difficult to predict. As the number of correspondencies is very large, it is necessary to reduce the search space by including constraints like scene knowledge.

In contrast to the two other presented matching techniques (*raster-based matching* and *feature-based matching*) within *relational matching* the similarity measure is a global one, including the whole image information (primitives, their attributes and the relations).

More details about *relational matching* can be found in (Förstner, 1995).

7.5 Summary

To find corresponding object points in different images (for a subsequent deformation analysis) we have described various approaches, namely the *raster-based matching*, the *feature-based matching* and the *relational matching*. We have implemented a matching technique, which is based on a combination of cross correlation (*raster-based matching*) and point detection. A comparison of matching techniques can be found in (Förstner, 1995).

With this chapter we have completed the description of the theoretical background necessary for our system. This theoretical knowledge was implemented in a concrete implementation, which we will present in the following chapter. Furthermore in Chapter 9 we will show some complex examples including the three main steps of our system: the image preprocessing and image enhancement, the point finding and the point filtering.

Additionally we will discuss an alternative technique to a knowledge-based decision system, the so-called *neural networks* and some concrete examples done in the course of development experiments.

Part IV

Implementation, Experiments and Alternative Technique

Chapter 8

Implementation

In the preceding chapters we have presented knowledge-based decision methods for image preprocessing and image enhancement, for finding *interest points* and for point filtering. The techniques described have been implemented in a common program system. The architecture of this implementation, the data format and the data base interface will be content of this chapter.

8.1 System Architecture

The developed program system (we have called it *"KBVMS"* = "Knowledge-Based Videometric Measurement System") was written in two different program languages. The knowledge-based system was implemented in CLIPS, a productive development tool which provides a complete environment for the construction of *rule- and object-based systems* (more details about CLIPS in Section 4.3). The rest of the implementation (image analysis, image processing, *interest operators*, user interface, and others) was realised in C++. The system architecture is shown in Figure 8.1.

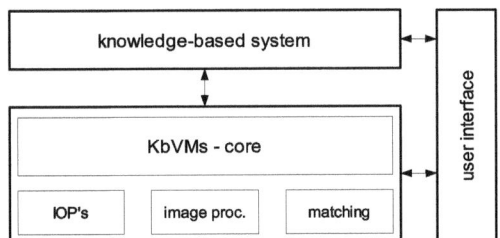

Figure 8.1: Architecture of the developed program system.

The program system was mainly developed to integrate all necessary program modules into one common user interface. This makes tests and experiments easier and permits "non-experts" to handle the system.

The KBVMS core contains the complete program and data base management. The communication between core component, different program modules and knowledge-based systems occurs by means of specified interface files.

8.1.1 Interface Files, Output Files and Data Flow

On the basis of the presented concept of the developed measurement system (Figure 6.14), we will describe in the following the structure of used file types.

For better understanding the data flow is shown in Figure 8.2.

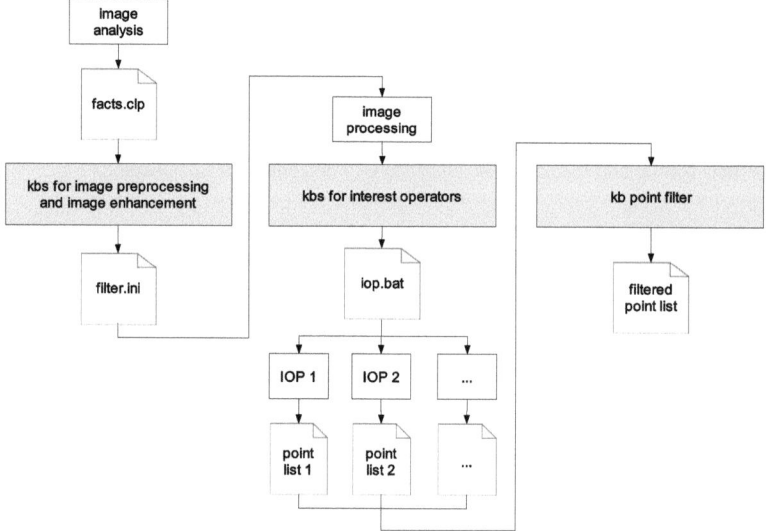

Figure 8.2: Data flow of the developed program system.

After having loaded an image (or captured it in an on-line system) the image analysis has to be done. The image features received in this way are stored in the "facts.clp" file, which has the following structure:

```
(Stat_Moments (lNr 0) (M1 value) (M1_f fuzzy_value) (M2 value) ...)
(1_Haralick_Moment (lNr 1) (H1_0 value) (H1_0_f fuzzy_value) (H1_45 value)...)
(2_Haralick_Moment (lNr 2) (H2_0 value) (H2_0_f fuzzy_value) (H2_45 value)...)
...
...
(UQ (lNr 14) (FT fuzzy_value) (RFL fuzzy_value) (S fuzzy_value))
```

The chosen syntax for "facts.clp" allows the loading of the image features into the knowledge base

8.1 System Architecture

as facts by means of (`load-facts "facts.clp"`). Facts are one of the basic forms for representing information in a CLIPS system. Each fact (each line in "facts.clp") represents a piece of information which has been placed in the current list of facts, called the fact list. Facts are the fundamental unit of data used by the rules (CLIPS, 2004).

All decisions are based on the extracted image features and therefore on the facts stored in the "facts.clp" file.

The result of the knowledge-based image preprocessing and image enhancement system is an output file ("filter.ini") which contains the relevant processing steps and their parameters. This file has the following structure:

```
[filter0]
FilterType=name of processing step 1
param=relevant parameter
[filter1]
FilterType=name of processing step 2
param=relevant parameter
...
...
```

The "filter.ini" file consists of groups of three lines. Each first line of a group represents a consecutive number combined with the word "filter", each second line lists the image preprocessing or image enhancement algorithm ("FilterType"), and the last line specifies the necessary parameters for the appropriate algorithm.

Based on the "filter.ini" file, the image preprocessing and image enhancement steps will be applied. This process results in an image better suited for the subsequent detection of *interest points*. After having processed the image the "facts.clp" file will be recreated. This is necessary since by image preprocessing and image enhancement the image features have been changed.

The knowledge-based point detection is based on the new created file "facts.clp" and results directly in a batch-file ("iop.bat"), which contains the relevant commands for the application of the *interest operators* including their parameter.

Each applied *interest operator* provides a list with the detected *interest points*. These point lists have a uniform format, which is as follows:

```
IOP_name   index   parameter1   x-coordinate   y-coordinate   parameter2   parameter3
...
```

The columns of the point file have the following meaning:

- `IOP_name`: name of the corresponding *interest operator* (possible values: F1, F2, HRS, HFVM);
- `index`: consecutive number beginning at 0;
- `param1`: properties parameters obtained from the Förstner 1, the Harris and the HFVM operator;
- `x-coordinate` and `y-coordinate`: pixel co-ordinates of the detected points;

- `param2` and `param3`: properties parameters obtained from the Förstner 2 operator.

The knowledge-based point filter unites these single lists and weights the *interest points*. The output file corresponds to the shown filestructure with an additional column which contains the weights of the point.

8.1.2 Knowledge Representation

As we explained in Section 4.2.1 we have implemented our knowledge base as *rule-based / object-oriented* approach. Such systems consist of two main components:

- a set of rules,
- a working memory.

This section gives a brief overview of the programming elements of the CLIPS programming language including the CLIPS OBJECT-ORIENTED LANGUAGE (COOL).

The syntax of a rule definition in CLIPS is very similar to the "pseudo-code" listed in Section 4.2.1. The difference is that the whole rule is enclosed in parentheses, as well as the preconditions and the actions. In the following a concrete example rule:

```
(
  defrule histo
    (declare (salience 200))
    (Stat_Moments (M1_f very_low | low | very_high | high))
    (Stat_Moments (M3_f low_positive | very_low_positive |
                        low_negative | very_low_negative))
    =>
    (assert (condition (histo yes)))
)
```

The `salience` rule property allows the user to assign a priority to a rule. If more than one rule has to be fired, the rule with the highest priority will be applied first.

We have divided the whole knowledge base (rule base) into three separated parts:

- knowledge base for the knowledge-based image preprocessing and image enhancement system;
- knowledge base for the knowledge-based point detection;
- knowledge base for the knowledge-based point filtering.

Each of these knowledge bases is stored in a separate file. This makes modification and extensions easy. A listing of the complete knowledge base can be found in Appendix C.1.

The second component of a *rule-based / object-oriented* approach is the *working memory*, a collection of *working memory elements*. These are instantiations of a *working memory type* (WMT) and can be considered as `record` declarations in PASCAL or `struct` declarations in C.

In CLIPS *working memory types* are realised as so-called `deftemplates`. Deftemplates provide the user with the ability to abstract the structure of a fact by assigning names to each field found within the fact. The `deftemplate` construct is used to create a template which can then be used by the facts. The following is an example for an implemented deftemplate:

8.1 System Architecture

```
(
  deftemplate Stat_Moments
    (slot lNr (type INTEGER))
    (slot M1 (type FLOAT))
    (slot M1_f (type SYMBOL)
           (allowed-symbols very_low low middle high very_high))
    (slot M2 (type FLOAT))
    (slot M2_f (type SYMBOL)
           (allowed-symbols very_low low middle high very_high))
    (slot M3 (type FLOAT))
    (slot M3_f (type SYMBOL)
           (allowed-symbols very_low_positive low_positive middle_positive
                            high_positive very_high_positive very_low_negative
                            low_negative middle_negative high_negative
                            very_high_negative))
)
```

Stat_Moments is a *working memory type* consisting of seven *slots*, namely lNr, M1 (mean), M1_f (fuzzy value of mean), M2 (variance), M2_f (fuzzy value of variance), M3 (skew) and M3_f (fuzzy value of skew) respectively. The type of each slot is INTEGER, FLOAT or SYMBOL. SYMBOL means that a symbol can be stored in the slots.

The symbols allowed for each of the slots are defined with "allowed-symbols". Type checking is performed during runtime in order to guarantee that the content of a slot satisfies its definition.

In our system there exist *deftemplates* definitions for histogram features (Stat_Moments), for each Haralick feature (1_Haralick_Moment, 2_Haralick_Moment, etc.) and for the image preprocessing and *interest operator* conditions (histo, thresholding, etc.).

The program structure described makes it possible to load and execute the single knowledge-based systems by means of the following batch commands, which are stored in a suitable batch file (for this example "start_rulebase_image_preprocessing.bat"):

```
(deffunction app-on-init())
(load "rulebase_image_preprocessing.clp")
(load-facts "facts.clp")
```

The app-on-init command initializes the application. The second command loads the constructs stored in the specified file into the environment. In our example this is the rule base for the knowledge-based image preprocessing and image enhancement system. The last function will assert a file of information as facts into the CLIPS fact list.

WXCLIPS was chosen as the development environment for the knowledge-based systems. WXCLIPS was developed to enable CLIPS programmers to write portable, graphical programs which run under several platforms. It is an extension to CLIPS and was developed by Julian Smart of the *Artificial Intelligence Applications Institute, University of Edinburgh* (wxClips, 2004). We have chosen WX-CLIPS for two reasons:

- WXCLIPS represents a library of CLIPS functions to access the wxWindows functionality,

- WXCLIPS is a simple stand-alone development environment for developing wxWindows applications using the extra CLIPS functions.

We have used the stand-alone development environment without the special WXCLIPS library. The knowledge-based systems (e.g. the part for the knowledge-based image preprocessing and image enhancement) will be called from the C++ main program by:

```
if ((int)ShellExecute(Handle,NULL,"wxclips.exe","-clips
                      start_rulebase_imgage_preprocessing.bat -start",
                      NULL ,SW_NORMAL)<32)
```

In the following section more details about the practical implementation of the developed prototype will be given.

8.2 The Prototype

The developed prototype *KBVMS* (**K**nowledge-**B**ased **V**ideometric **M**easurement **S**ystem) runs under Microsoft Windows 2000 and integrates the described program modules into a common user interface.

As we mentioned above the knowledge-based systems have been developed in CLIPS / WXCLIPS. The other program modules have been realised by means of the Borland C++ Builder (Version 5).

The graphical user interface (GUI) of the prototype was realised in order to free the user from learning a complex command languages. The complete process chain should have been integrated into a logical handling sequence by arranging all GUI elements on the prototype main window.

This main window (shown in Figure 8.3) is subdivided into three areas:

(a) a part which contains the GUI elements for process controlling;

(b) a part which shows program outputs, like the calculated image features and others;

(c) a part which contains the GUI elements for additional commands.

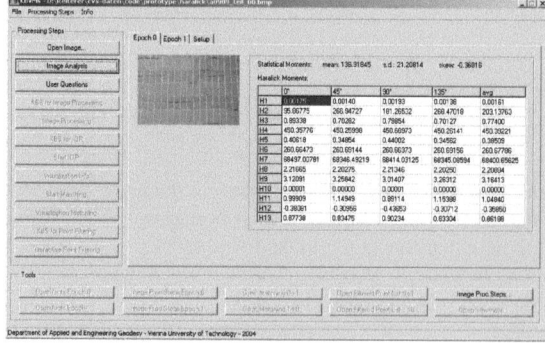

Figure 8.3: Main window of the developed prototype *KBVMS*.

8.2 The Prototype

For the user-queries described in Section 3.4, we have designed a dialog box. It will be used to obtain the user input needed to complete the image analysis. This dialog box is shown in Figure 8.4. For the first user question (*UQ1: What kind of type is the facade?*) an expressive example image will be shown.

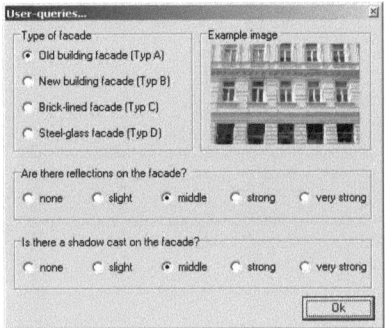

Figure 8.4: Dialog box for user-queries.

The preprocessed image and the detected *interest points* are shown on the right part of the main window, which contains all output data. The small picture can be enlarged. After the knowledge-based point filtering in this type of enlarged image (shown in a separate window) the interactive point filtering can be done.

As we explained in Section 6.5.2, in a final step the user has the possibility to let only points with the same weight be shown and to remove point groups which are weighted differently. This is done with a slider in the top left corner of the window.

In conclusion it can be said that the prototype developed includes all the techniques and methods developed in the course of the work at hand. Single tests and complete simulations could be carried out efficiently and fast.

Having described the single knowledge-based systems (image preprocessing and image enhancement, point detection, point filtering) only examples for these sub-problems have been presented, without the complete process chain. Complete simulations including the whole process will be presented in the next chapter.

Chapter 9

Experiments

In this chapter we will present the complete functionality of the developed prototype-system by two extended examples. As the developed measurement system is focused on monitoring of building facades, in the following examples this object type will be concerned, too. In the previous chapters we have only presented examples, which demonstrated single process results, but never the whole process sequence including its results. The latter will be done in the following two sections.

9.1 First Example

The first example shows a highly underexposed (too dark) image (Figure 9.1a). The histogram (Figure 9.1b) shows that all grey-level values are less than 100 and concentrated as a narrow peak (low contrast). First of all image analysis has to be done. The resulting image features are listed in Table 9.1.

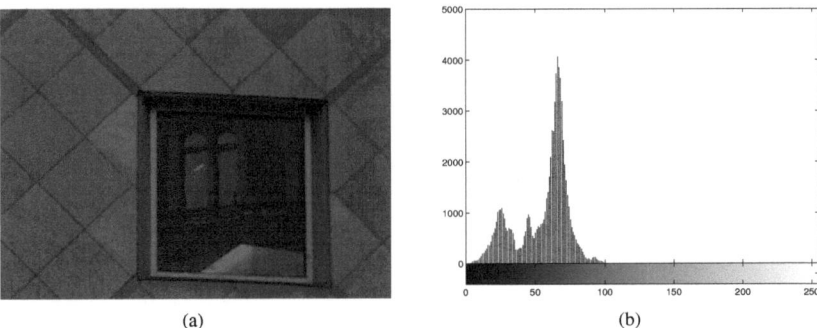

Figure 9.1: The underexposed image and its histogram.

The user questions (UQ) (see Section 3.4) have been answered as follows:
- UQ1: What kind of type is the facade? → new building facade (= type B).
- UQ2: Are there any reflections on the facade? → middle.

M_1	M_1 - f.v.	M_2	M_2 - f.v.	M_3	M_3 - f.v.				
55.6004	very low	18.3897	very low	-0.7580	very high negative				
	0°	f.v.	45°	f.v.	90°	f.v.	135°	f.v.	average	f.v.
H_1	0.0035	high	0.0029	high	0.0038	high	0.0030	high	0.0033	high
H_2	21.0878	v.low	37.7307	v.low	20.6355	v.low	38.2739	v.low	29.4319	v.low
H_5	0.4426	high	0.3743	mid.	0.4632	high	0.3826	mid.	0.4157	high
H_9	2.7974	low	2.9185	low	2.7810	low	2.9086	low	2.8514	low

Table 9.1: Extracted image features for Figure 9.1a.

- UQ3: Is there a shadow cast on the facade? → none.

Due to the calculated image features the knowledge-based system (see rule base in Section 5.3) chooses *edge detection* as one of the suitable image preprocessing methods and therefore the system gives the user the possibility to overrule this decision. If the user makes use of this, *edge detection* will not be applied; *median filtering* and *grey-level scaling* (*image brightening*) will remain. The processed image is shown in Figure 9.2a.

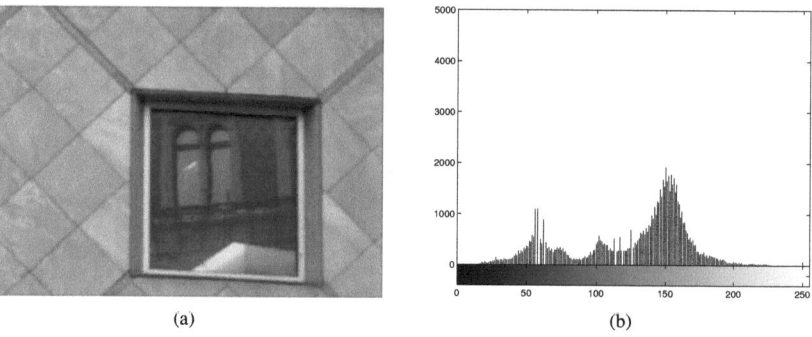

(a)   (b)

Figure 9.2: The image from Figure 9.1 after image preprocessing / enhancement and its histogram.

The application of *grey-level scaling* results in a brighter image with increased contrast; the *median filter* has reduced the noise by smoothing the image. Useful details, like edges and corners, are now (after image preprocessing and image enhancement) visible. The histogram of the resulting image (Figure 9.2b) is more spread and the grey-level values are distributed more regular.

If the user does not overrule the system decision, *edge detection* will have to be carried out. As we have shown in Section 5.4, *edge detection* (together with *gauss filtering* as a preprocessing method and *thresholding* as a postprocessing method) reduces the information content of the image substantially. Grey-level values are lost and only places with strong intensity contrast are conserved. For this reason and due to the high quality of the processed image without *edge detection* (Figure 9.2) this case will not be contemplated in the following.

After image preprocessing and image enhancement the image properties have changed, so that, before the knowledge-based system chooses a suitable algorithm (*interest operator*) for finding *interest points*, image analysis has to be done again. The resulting image features are listed in Table 9.2.

9.1 First Example

M_1	M_1 - f.v.	M_2	M_2 - f.v.	M_3	M_3 - f.v.				
126.9833	middle	41.3374	middle	-0.7607	high negative				
	0°	f.v.	45°	f.v.	90°	f.v.	135°	f.v.	average	f.v.
H_1	0.0011	mid.	0.0009	mid.	0.0013	mid.	0.0009	mid.	0.0011	mid.
H_2	68.0373	low	126.9122	mid.	61.9111	low	126.4453	mid.	95.8265	low
H_5	0.3630	mid.	0.2781	low	0.3830	mid.	0.2876	low	0.3279	mid.
H_9	3.3050	mid.	3.4666	mid.	3.2829	mid.	3.4488	mid.	3.3758	mid.

Table 9.2: Extracted image features for Figure 9.2a.

If we compare the preconditions of the rules described (Section 6.2) with the extracted image features from Table 9.2, the following interest operators are selected: Förstner 2 and the Harris operator.

Figure 9.3a shows the applied *interest operators*: Förstner 2 with $q_{min} = 0.2$, $W_{min} = 360$ and $R = 3$; the Harris with $\sigma = 1.0$, $N_0 = -0.04$ and $corn_{min} = 0.0018$. The parameters have been chosen on the basis of the rules defined in Section 6.3.3.

(a)   (b)

Figure 9.3: (a) *Interest points* detected with the Förstner 2 and the Harris operator; (b) *interest points* after rule-based point filtering

In Figure 9.3a it can be seen that *interest points* are generally detected on the regular structure of the facade, only a small number of isolated single points are detected inside these "structure lines". These points result from local grey-level differences, like "fault-pixels". A more problematic area is the glass window, where many *interest points* are caused by reflections. Changes of parameter values would remove the undesirable points on the glass windows, as the desired *interest points*, too (the grey-level differences in this area are the same as those of the "structure lines" of the facade). Undesirable points can only be eliminated (removing is realised by means of weighting) by a suitable point filtering technique.

As described in Section 6.5 the developed point filter works by means of two methods: on basis of defined rules and interactive. The first method uses two criterions for weighting each point: (1) how many *interest operators* detect the point and (2) which "property-parameters", obtained from the corresponding *interest operator*, the point has. In this example no points have been detected by more than one *interest operator*. Therefore, in the first step the weights remain unchanged (*weight* = 1). In the second step the returned values by the *interest operators* are inspected (details about this step can be found in Section 6.5). Figure 9.3b shows the points after this filtering process. It can be seen that too many points have been filtered. Only a small number of *interest points* beside the window have been preserved. Additionally points inside the glass window are still present. These two deficiencies can be removed by means of the interactive part of filtering process. The user can remove the points on the glass by drawing a rectangular window in the graphical output. Points inside these selected windows will be removed (weighted with -1).

For the final output the user has the possibility to let display only points with the same weight and to remove point groups which are weighted differently. In our example this step can be used to let reappear the filtered *interest points*. The resulting *interest points* are shown in Figure 9.4a.

To show that a knowledge-based decision system is an improvement for the whole process sequence we have put the system without the knowledge-based components on an unpractised user disposal. The result is shown in Figure 9.4b. The application of *interest operators* (Förstner 2 and the Harris operator) has been done without image preprocessing and image enhancement (Förstner 2 with $q_{min} = 0.2$, $W_{min} = 2000$ and $R = 3$; the Harris with $\sigma = 1.0$, $N_0 = -0.04$ and $corn_{min} = 0.1$).

(a)   (b)

Figure 9.4: (a) Final result after image preprocessing and image enhancement, application of *interest operators* and point filtering; (b) *interest points* detected without a knowledge-based decision system (done by an unpractised user).

9.2 Second Example

The second example presents an overexposed (too bright) image with extremely low contrast (Figure 9.5a). The histogram (Figure 9.5b) shows that not all grey-level values are present. Some of the grey-levels show zero values in the histogram, indicating that none of the pixels have those values. First of all image analysis has to be done. The resulting image features are listed in Table 9.3.

The user questions (UQ) (see Section 3.4) have been answered as follows:

- UQ1: What kind of type is the facade? → steel-glass (= type D).
- UQ2: Are there any reflections on the facade? → none.
- UQ3: Is there a shadow cast on the facade? → none.

9.2 Second Example

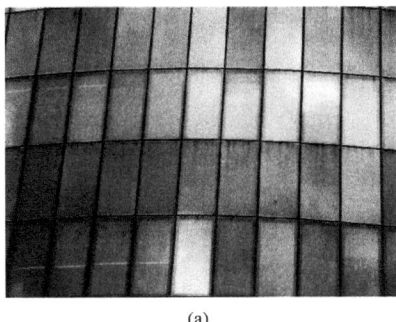 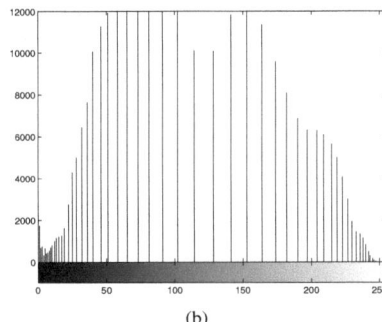

(a) (b)

Figure 9.5: The overexposed image and its histogram.

M_1	M_1 - f.v.	M_2	M_2 - f.v.	M_3	M_3 - f.v.
130.7274	middle	72.0389	high	-0.0471	very low negative

	0°	f.v.	45°	f.v.	90°	f.v.	135°	f.v.	average	f.v.
H_1	0.0121	v.high	0.0093	v.high	0.0139	v.high	0.0091	v.high	0.0111	v.high
H_2	9.5635	v.low	14.2014	v.low	5.9788	v.low	15.3021	v.low	11.2614	v.low
H_5	0.6757	v.high	0.6017	v.high.	0.7253	v.high	0.5940	high	0.6492	v.high
H_9	2.2073	low	2.2997	low	2.1173	low	2.3120	low	2.2341	low

Table 9.3: Extracted image features for Figure 9.5a.

If we compare the preconditions of the described rules (Section 5.3) with the extracted image features from Table 9.1 the following image preprocessing and image enhancement algorithms have to be done: *histogram equalization* and *median filtering*. The application of *median filtering* depends on user interaction. Contrast expansion (in our case by means of *histogram equalization*) is mostly accompanied by an increasing visibility for noise. Due to this fact, *median filtering* is advisable. *Histogram equalization* has reassigned the grey values, so that the visual contrast for the image has been increased. *Median filter* has smoothed the image and therefore reduced the noise. The resulting image is shown in Figure 9.6.

A comparison between the image before (Figure 9.5) and after image preprocessing and image enhancement (Figure 9.6) shows that useful image details are visible, like the steel-glass structure on the left side of the image. The *histogram equalization* has brightened the image so that the left part of the image presents more contrast, and the right part has not been overexposed.

After image preprocessing and image enhancement have been carried out, the image features have been changed. Therefore, image analysis has had to be done on the processed image (Figure 9.6a). The resulting image features are listed in Table 9.4.

On the basis of these image features, the knowledge-based system (Section 6.2) chooses a suitable *interest operator* and its parameters. For this example the following *interest operators* have been selected: Hierarchical Feature Vector Matching (HFVM) and the Harris operator.

Figure 9.7a shows the applied *interest operators*: HFVM operator with the feature set defined in Equation 6.6; the Harris with $\sigma = 1.0$, $N_0 = -0.04$ and $corn_{min} = 0.000085$. The parameters have

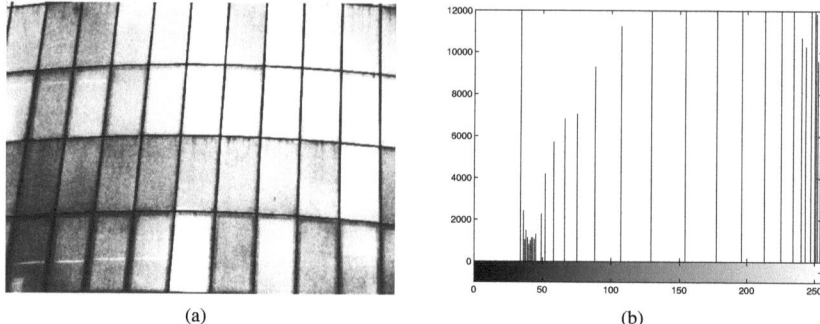

Figure 9.6: The image from Figure 9.5 after image preprocessing / enhancement and its histogram.

M_1	M_1 - f.v.	M_2	M_2 - f.v.	M_3	M_3 - f.v.				
183.5020	high	77.7008	high	-0.7871	high negative				
	0°	f.v.	45°	f.v.	90°	f.v.	135°	f.v.	average	f.v.
H_1	0.0296	v.high	0.0264	v.high	0.0316	v.high	0.0261	v.high	0.0284	v.high
H_2	9.2827	v.low	13.7174	v.low	5.6948	v.low	14.8105	v.low	10.8763	v.low
H_5	0.7168	v.high	0.6534	v.high	0.7645	v.high	0.6452	v.high	0.6950	v.high
H_9	1.9772	v.low	2.0583	low	1.8905	v.low	2.0707	low	1.9991	v.low

Table 9.4: Extracted image features for Figure 9.6a.

been chosen on the basis of the rules defined in Section 6.3.3. It can be seen that points are not only detected on the steel structure of the facade, but also inside the glass windows. The latter points result from strong local grey-level differences on these areas. First of all the left part of the image represents such problem zones.

As we have shown in Section 6.4 changing the chosen parameters can not eliminate this problem. Undesirable points can only be eliminated by a filtering technique.

As in *Example 1* described the developed point filter works *rule-based* and by means of an interactive process between user and system. The first step results in a point cloud shown in Figure 9.7b. The points have been weighted differently depending on the number of *interest operators*, which detect the same points (in our case the weights remain unchanged, since no points have been detected by more than one *interest operator*) and depending on the returned values by the *interest operators* (details about this step can be found in Section 6.5).

It can be seen that most of the points inside the window areas have been filtered. Most of the preserved *interest points* represent the steel structure of the facade and will be necessary for subsequent processing steps, like deformation analysis or object reconstruction. Only for the image part with low contrast (on the bottom left) no *interest points* have been detected. This is caused by the requirement of a minimal local grey-level difference for the correct working of *interest operators*.

The user has the possibility to remove these points on the glass by drawing a rectangular window in the graphical output. Points inside these selected windows will be removed (weighted with -1). The final output is shown in Figure 9.8a.

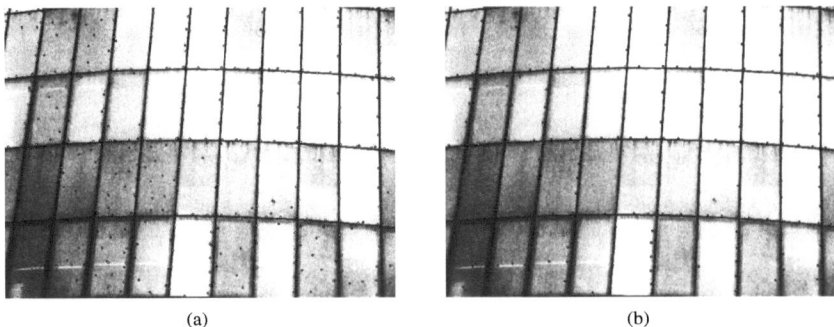

Figure 9.7: (a) *Interest points* detected with the HFVM and the Harris operator; (b) *interest points* after rule-based point filtering.

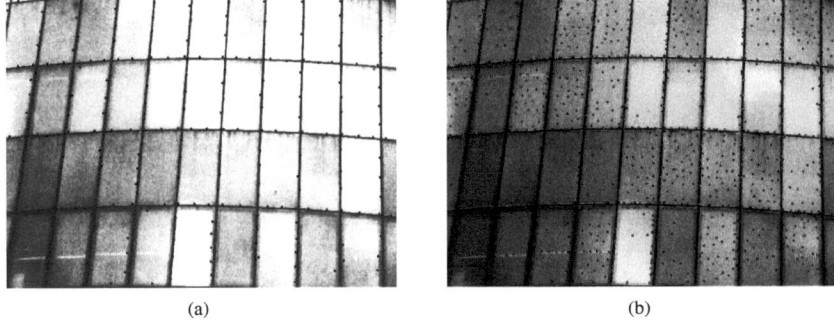

Figure 9.8: (a) Final result after image preprocessing and image enhancement, application of *interest operators* and point filtering; (b) *interest points* detected without a knowledge-based decision system (done by an unpractised user).

Like in *Example 1*, we have put the system without the knowledge-based components on an unpractised user disposal, to show the improvement by the developed system. The result is shown in Figure 9.8b. The user has applied the Förstner 2 and the Harris operator without image preprocessing and image enhancement (Förstner 2 with $q_{min} = 0.2$, $W_{min} = 10000$ and $R = 3$; the Harris with $\sigma = 1.0$, $N_0 = -0.04$ and $corn_{min} = 0.0053$).

9.3 Summary

In this chapter we have presented two examples, which show the complete process sequence: image preprocessing/image enhancement, point detection and point filtering. The examples have shown that the developed knowledge-based decision system is a considerable improvement over a conventional system without integrated domain-knowledge. This was shown by a comparison of resulting *interest points* detected by our developed system on the one hand, and detected by the same system but without knowledge-based components on the other hand. In spite of automation of the whole process, the user is the last instance to finalize the sequence and to confirm the point list.

The two presented examples are part of an extensive test program done with the developed system on about 120 images. In all test sceneries the process results in a final point list, which is suitable for the provided processing steps (deformation analysis or object reconstruction). A reduced usability for the system exists for images or image parts with very low contrast, since an *interest operator* needs a minimal local grey-level difference to work correctly.

Chapter 10

Alternative Technique – Artificial Neural Networks

As we have explained in Section 3.5, various approaches can be used for a decision system. We have realised our decision support system as knowledge-based approach. An other popular method are the so-called artificial neural networks, presented in this chapter. At the beginning we give a short introduction into artificial neural networks, which should not be seen as complete treatise of this subject.

10.1 Introduction

Artificial neural networks are based on a different approach of problem solving in comparison with conventional computer software. Conventional software use an algorithmic approach; the program follows a set of instructions in order to solve the problem. A condition to this is that the specific problem solving steps are known. The disadvantage of such program systems is that the problem solving capability is restricted to problems that we already understand, from which we know how to solve them and to problems for which we can formulate a solving algorithm.

Artificial neural networks process information in a similar way the human brain does. The network consists of a set of highly interconnected processing elements, the so-called *neurons*. Artificial neural networks cannot be programmed to solve a specific problem; they learn by examples. The disadvantage of such an approach is that the network finds out how to solve the problem by itself, therefore the user receives no declaration how the problem was solved. Furthermore neural networks operations can be unpredictable.

The most basic components of artificial neural networks are modeled after the structure of the human brain. Therefore artificial neural networks have a strong similarity to the (biological) brain and most of the terminology is borrowed from neuroscience. In the following we will explain some basics of the human brain and compare it with an artificial approach.

10.1.1 The Neuron

The biological neuron: The basic element of the human brain is a specific type of cell, the so-called *neuron*. Each *neuron* uses biochemical reactions to receive, process and transmit information. The human brain is a collection of about 10 billion interconnected neurons. The power of the brain comes from the number of these basic components and the multiple connections between them (each *neuron* can be connected with up to 200000 other neurons).

All natural *neurons* have four basic components: dendrites, soma, axon, and synapses (Figure 10.1 shows a simplified biological neuron). The dentrites accept the input, the soma processes the input and the axon turns the processed input into outputs. The synapses represent the electrochemical contact between *neurons*.

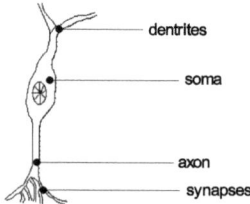

Figure 10.1: The biological neuron.

The artificial neuron: As complicated as the biological neuron is, it may be simulated by a very simple model. An artificial neuron is a device with many inputs and one output (Figure 10.2). Each input is multiplied by a weight. In the simplest case, these products are summed, fed through a transfer function to generate a result, and then put out. The neuron has a threshold value. If the sum of all the weights of all active inputs is greater than the threshold, the neuron will be active.

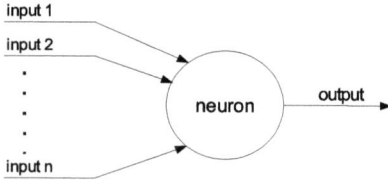

Figure 10.2: A simple artificial neuron.

10.1.2 Artificial Neural Network:

Artificial neural networks are a simple clustering of artificial neurons. These networks are subdivided into layers, which are then connected among each other. Basically, all artificial neural networks have a similar structure and consist of three types of layers:

- input layer represents the first interface to the real world (to receive the inputs);

- output layer represents the second interface to the real world (to provide the networks outputs);
- hidden layer represents the rest of the network (to transform the information from the input layer to the output layer).

The single artificial neurons are connected (normally unidirectional) via a network of paths. Each neuron receives inputs from many other neurons, but produces a single output, which is communicated to other neurons.

There exist different types of connections between neurons (fully connected, partially connected and others). We use a *fully connected feedforward network*, by which each neuron on the first layer is connected to every neuron on the second layer. The neurons on the first layer send their output to the neurons on the second layer, but they do not receive any input back form the neurons on the second layer. A simple form of a *fully connected feedforward network* is shown in Figure 10.3.

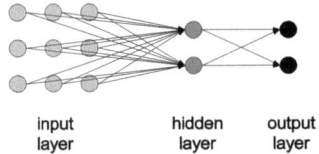

Figure 10.3: A simple form of a *fully connected network*.

The artificial neural network has, in its simplest form, two modes of operation: (a) the training mode and (b) the using mode. In the *training mode*, the neurons can be trained to fire (or not) for particular input patterns. This "learning" results in changes of the connection weights. The strength of each connection between the neurons is stored as a weight-value for the specific connection. The system learns new knowledge by adjusting these connection weights.

In the using mode, any (possible new) input patterns are detected at the input. The artificial neural network solves the problem (determines the output) on the basis of trained connection weights.

More detail about artificial neural networks and their application can be found in (Zell, 1997).

In order to investigate the usability of a neural network for some of our problems, we performed two experiments described in the next section.

10.2 Experiments

We have developed and tested artificial neural networks for the following two problems:

- choice of a suitable *interest operator*;
- classification of facade types (so far done by the user by means of a user question, described in Section 3.4).

The first problem was chosen, since it is one of the core problems of our process sequence (the knowledge-based decision system for the choice of a suitable *interest operator* is described in Chapter 6).

The second problem was chosen, since image classification is one of the most typical problems for the application of artificial neural networks. An automatic classification tool, which is able to classify facades dependably, could substitute the first user question (UQ1) described in Section 3.4 and would be another important step to an autonomous measurement system.

It is notable that all experiments described in the following have been done with the SNNS (Stuttgart Neural Network Simulator) simulator. SNNS is a software simulator for neural networks developed at the Institute for Parallel and Distributed High Performance Systems (IPVR) at the University of Stuttgart (SNNS, 2004).

10.2.1 Artificial Neural Network for the Choice of a Suitable *Interest Operator*

This first artificial neural network has the goal to choose a suitable *interest operator*. For the sake of simplicity, this choice was reduced to two *interest operators* without their parameters. The developed network uses the grey level value of each pixel in the image as input. The images have been resized to 64×48 pixels (input layer). This size reduced the dimensionality of the input object to a reasonable quantity, while still keeping intact the visual content of the object. For output the Förstner 2 and Harris operator have been chosen (output layer).

The number of hidden-units is directly related to the capabilities of the network. For the best network performance an optimal number of hidden-units must be properly determined. In literature for similar problems a hidden layer of 20×20 nodes has been created (Carpenter, 1992; Zell, 1997). Therefore we have built a hidden layer of 20×20 nodes, too.

Figure 10.4 shows the developed network. It was trained using 100 images; 50 images for each *interest operator*. The test set consisted of grey values of the 100 images and the condition values for *interest operators*; e.g. if a facade is suited for the Förstner 2 operator, the condition values will be 1 for the Förstner 2 node and 0 for the Harris node.

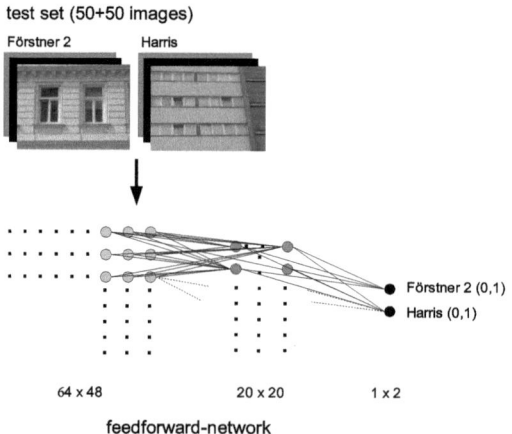

Figure 10.4: Artificial neural network for the choice of a suitable *interest operator*.

10.2 Experiments

The network reached a set of weights capable of correctly identifying the training images after 200 iterations. For testing we used 10 new unknown images (5 images for each *interest operator*) as input data.

The network has correctly classified 50% of the data. It has chosen the Förstner 2 operator for all testing images. If the testing set is modified, such that all testing images are suited for the Förstner 2 operator, we will get 100% of correctly classified cases, at the other hand, a modification to testing images, which are suited for the Harris operator, provides 0% of correctly classified cases. This result is inadequate. Additional training (made for up to 2000 iterations) of the network, variations of the dimensions (up to an input layer of 128×96 and a hidden layer of 40×40 nodes) had no significant effect on its ability to correctly choose *interest operators*, but increased the training times.

The result of this experiment shows that a reliable choice of suitable *interest operators* by means of the developed artificial neural network is unfeasible, at least not by a straightforward means.

10.2.2 Artificial Neural Network for the Classification of Facade Types

The goal of this artificial neural network is to classify facades into the four types defined in Section 3.4. The network takes an image as input and places it in a certain category as output. As in *Example 1* the developed network uses the grey level value of each pixel in the image as input. The images have been resized to 64×48 pixels (input layer). For output the four facade types have been chosen (output layer). The hidden layer was built from 20×20 nodes. Figure 10.5 shows the developed network.

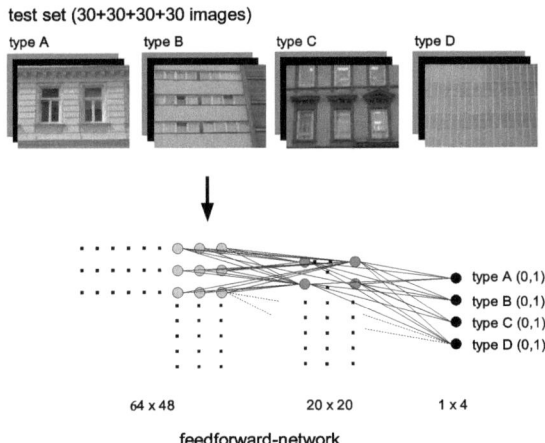

Figure 10.5: Artificial neural network for the classification of facade types.

The network was trained using 120 images with 30 images for each facade type. The network reached a set of weights capable of correctly identifying the training images after 400 iterations.

For testing we used 20 new unknown images (5 images for each facade type) as input data. The network has correctly classified 50% of the data. This is significantly better than a random result from the previous example. The low percentage results from the different nature of facades in practice; each building facade has its own visual characteristics.

Figure 10.6 shows two correctly and two falsely classified images for the facades type B (new building facade) and D (steel-glass facade).

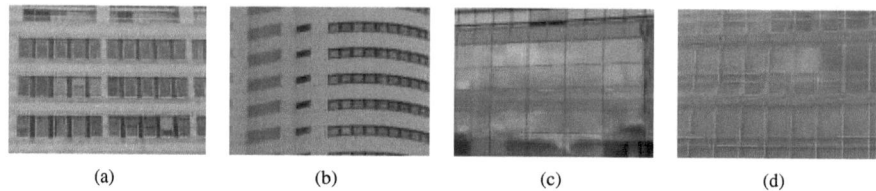

(a) (b) (c) (d)

Figure 10.6: (a) New building facade classified as steel-glass facade; (b) new building facade classified correctly; (c) steel-glass facade classified correctly; (d) steel-glass facade classified as new building facade.

Additional training (made for up to 2000 iterations) of the network, variations of the dimensions (up to an input layer of 128×96 and a hidden layer of 40×40 nodes) had no significant effect on its ability to correctly choose *interest operators*, but increased the training times. For 2000 iterations a training time of about 12 hours was necessary.

10.3 Summary

In this chapter we have presented an alternative technique to a knowledge-based approach: artificial neural networks. They process information in a similar way the human brain does. We have tested artificial neural networks to automate two process steps in our working chain: the choice of a suitable *interest operator*, and a classification of facade types (so far done by the user).

For the first problem (the choice of a suitable *interest operator*), an artificial neural network was unsuitable, the approach failed. For the second problem (the classification of facade types) we obtained better results; 50% of tested facades have been classified correctly. Maybe this percentage value could be increased by expanding the training data. Altogether, the insufficient results are caused by the different nature of facades in practice; each building facade has its own visual characteristics.

Our experiments have shown that the two problems are not solvable with a straightforward artificial neural network approach. In particular for the choice of a suitable interest operator we see the experiments as strong indicator that a satisfactory neural network solution, if one is feasible at all, will be difficult to find and involve high effort. On the other hand the experiments for the classification of the facades give a glimpse of hope that a neural network approach can be successful for this problem. A knowledge-based approach to solve this problem seems difficult.

Neural networks, knowledge-based systems and conventional algorithmic programs are not in competition, but complement each other. There are tasks more suited to an algorithmic approach like arithmetic operations, tasks that are more suited to neural networks (e.g. pattern recognition) or to a knowledge-based approach (e.g. complex decision systems). Even more, a large number of tasks requires systems that use a combination of these approaches in order to perform at maximum efficiency.

Part V

Conclusion

Chapter 11

Conclusion and Future Work

In this work, a new knowledge-based decision system for an on-line videotheodolite-based multi-sensor system has been described. The main task of our development has been the automation of different decision makings in the course of the measurement process. The decision support system, a specific class of computerized information system that supports the user by decision-making activities, has been realized with a knowledge-based approach. The separation of domain-knowledge from the reasoning mechanism, one of the fundamental concepts of such an approach, leads to the biggest advantages in comparison to conventional software: the system is easily modifiable and extensible.

To restrict the development process our measurement system has been focused on monitoring of building facades. This object type offers a wide range of structures and can be represented by simple line geometry. Integrated knowledge, examples and simulations have been fitted according to this.

The decision process is based on image features which represent the decisive image properties. For an on-line system a fast execution of such a feature extraction (image analysis) is necessary. In our system this is done by statistical feature extraction techniques.

In the course of our measurement process we have developed knowledge-based subsystems for the three core decisions:

- selection of suitable image preprocessing and image enhancement algorithms,
- selection of suitable *interest operators* and
- selection of points (point filtering) which are convenient for subsequent process steps (e.g. deformation analysis).

The algorithms, their parameters and their execution order are selected on the basis of the extracted statistical image features by the developed knowledge-based system.

Additionally a matching routine has been developed to find corresponding points in two or more measurement epochs. The developed method is based on a combination of cross correlation to describe the local similarities and *interest points* to specify the co-ordinates.

The knowledge-based components have been embedded in a prototype system, to provide the described program modules under a common user interface.

The developed system presents a basic approach for an automated on-line videotheodolite-based multisensor system. In such a system, the degree of automation can be very high, whereas by decision-making, human interaction remains an important part of the workflow even though the amount of decisions done by the user can be reduced considerably to a minimum. The system is of limited usability for images or image parts with very low contrast, since an *interest operator* needs a minimal local grey-level difference to work.

There are still many possibilities to improve the operability of the system. The most interesting and important ones for the future are:

- Improvement of the developed image preprocessing and image enhancement system, by integrating more algorithms, like *canny edge detection* or the *Wiener noise reduction filter*.
- The implemented knowledge-based point detection process could be improved by expanding the usability of the *interest operators*, first of all of the Hierarchical Feature Vector Matching operator. It would be desirable to extend this operator by several feature sets (the current implementation has only one), which could be selected depending on the points to be detected. Also an on-line generation of feature sets is possible and envisaged for future extensions of the system.
- The developed point filter has a lot of potential for an improvement of the system. Rules which do not only filter single points (by means of threshold values), but also regard the constellation of *interest points*, like filtering of points inside a glass window, could be implemented.
- The described system was only tested in a simulated off-line process. Therefore an important development step is the implementation in an on-line measurement system.

Beside the improvement of the system, the degree of automation for the whole system should be increased by integrating other sensors in the measurement process. A suggestive extension could be the integration of 3D laser scanners. The data of the different sensors have to be merged by a special data fusion process, which could be knowledge-based. Such a system provides an immense number of 3D data, both from the videotheodolite system and from the laser scanner. This point cloud may be reduced by filtering, even if not very effective. A new approach could build on cognitive vision and could work in a two step mode:

(1) The captured data (image and laser scanner data) must be used to produce a description of the image. This can be done by recognizing and assigning different objects to proper categories together with information about the object and relevant parameters. This process results in a special kind of information system.

(2) The yield information is used to produce actions into the physical world, like measuring a special part of the object.

Such a new measurement system would benefit from the efficiency of the 3D laser scanner, from the image information captured by the videotheodolites, and from the automation of decision processes basing on cognitive vision. The result would be a (semi) automated measurement system which is able to act and react in a known environment to unknown situations.

Bibliography

Baker, Simon / Nayar, Shree: Global Measures of Coherence for Edge Detector Evaluation. In: Proceedings of the Conference on Computer Vision and Pattern Recognition, pp. 373-379, Colorado, USA, 1999.

Bässmann, Henning / Kreyss, Jutta: Bildverarbeitung ad Oculos. 3rd Edition, Springer Verlag, Berlin/Heidelberg/New York, 1998.

Beierle, Christoph / Kerner-Isberner, Gabriele: Methoden wissensbasierter Systeme. 1st Edition, Vieweg, Braunschweig/Wiesbaden, 2000.

Bortz, Jürgen: Statistik für Sozialwissenschaftler. 5th Edition, Springer, Berlin, 1999.

Bowyer, Kevin / Kranenburg, Christine / Dougherty, Sean: Edge Detector Evaluation Using Empirical ROC Curves. In: Proceedings of the Conference on Computer Vision and Pattern Recognition, pp. 354-359, Colorado, USA, 1999.

Brezing, Horst: Entwicklung eines Expertensystems zur wissensbasierten Deformationsanalyse. PhD Thesis, RWTH Aachen, 2000.

Brownston, Lee / Farrell, Robert / Kant, Elaine / Martin, Nancy: Programming Expert Systems in OPS5: An Introduction to Rule-Based Programming. Addison-Wesley, New York, 1985.

Carpenter, Gail A.: Neural networks for vision and image processing. MIT Press, Cambridge, Mass., 1992.

Chmelina, Klaus: Wissensbasierte Analyse von Verschiebungsdaten im Tunnelbau. PhD Thesis, Vienna University of Technology, 2002.

CLIPS-Project-page: http://www.ghg.net/clips/CLIPS.html, 2004.

Eiter, Thomas / Mannila, Heikki: Distance Measures for Point Sets and Their Computation. Acta Informatica, Volume 34/2, pp. 109-133, 1997.

Fabiankowitsch, Johannes: Automatische Richtungsmessung mit digitalen Differenzbildern. PhD Thesis, Vienna University of Technology, 1990.

Förstner, Wolfgang: Statistische Verfahren für die automatische Bildanalyse und ihre Bewertung bei der Objekterkennung und -vermessung. DGK, Volume C, No. 370, München, 1991.

Förstner, Wolfgang: Matching Strategies for Point Transfer. In: Proceedings of the Photogrammetric Week 95, Fritsch / Hobbie (Editors), Wichmann Verlag, Heidelberg, 1995.

Free Online Dictionary of Computing: http://dict.die.net, 2004.

Gonzalez, Rafael C. / Wintz, Paul: Digital Image Processing. 5th Edition, Addison-Wesley Publishing, London, 1982.

Gonzalez, Avelino J. / Dankel, Douglas D.: The Engineering of Knowledge-based Systems. 1st Edition, Prentice-Hall Inc., 1993.

Gottlob, Georg / Früwirt, Thomas / Horn, Werner: Expertensysteme. 1st Edition, Springer, Wien/New York, 1990.

Görz, Günther / Rollinger, Claus-Rainer / Schneeberger, Josef (Editors): Handbuch der künstlichen Intelligenz. 3rd Edition, Oldenbourg Publishing, München/Wien, 2000.

Haralick, Robert M. / Shapiro, Linda G.: Computer and Robot Vision. 1st Edition, Addison-Wesley Publishing, New York, 1993.

Harris, Chris / Stephens, Mike: A Combined Corner and Edge Detector. In: Proceedings of the 4th ALVEY Vision Conference, Matthews (Editor), University of Manchester, England, 1988.

Hitz, Martin: C++ Grundlagen und Programmierung. 1st Edition, Springer, Wien/New York, 1992.

Hovenbitzer, Michael: Zur Automation berührungsloser 3D-Objekterfassung im Nahbereich. PhD Thesis, University of Technology Darmstadt, 2001.

ILOG - White Paper: http://www.ilog.com, 2003.

Kahmen, Heribert / Niessner, Anton / De Seixas, Andrea: 3D Object Sensing Using Rotating CCD Cameras. In: Digital Image Analysis, Kropatsch / Bischof (Editors), 1st Edition, Springer, Berlin/Heidelberg/New York, 2001.

Kahmen, Heribert / Suhre, Horst: Ein lernfähiges tachymetrisches Vermessungssystem zur Überwachung kinematischer Vorgänge ohne Beobachter. Zeitschrift für Vermessungswesen (ZfV), Volume 08, pp. 345-351, DVW, 1983.

Kahmen, Heribert / Roic, Miodrag: A New Generation of Measurement Robots for Object Reconstruction without Targeting. In: Optical 3-D Measurement Techniques III, Grün / Kahmen (Editors), Herbert Wichmann Verlag, Karlsruhe, 1995.

Kuhlmann, H: An expert system for bridge monitoring. 7th International FIG-Symbosium on Deformation Measurements, Banff, Canada, 1993.

Mischke, Alfred: Entwicklung eines Videotheodolite-Messsystems zur automatischen Richtungsmessung von nicht signalisierten Objektpunkten. PhD Thesis, Vienna University of Technology, 1998.

Mischke, Alfred / Kahmen, Heribert: A New Kind of Measurement Robot System for Surveying of non Signalized Targets. In: Optical 3-D Measurement Techniques IV, Grün / Kahmen (Editors), Herbert Wichmann, Karlsruhe, 1997.

Moravec, Hans: Towards Automatic Visual Obstacle Avoidance. In: Proceedings of the International Joint Conference on Artificial Intelligence, p. 584, 1977.

Kolesnik, Marina / Paar, Gerhard / Bauer, Arnold / Ulm, Michael: Algorithmic Solution for Autonomous Vision-based Off-road Navigation. In: Proceedings of SPIE: Enhanced and Synthetic Vision, Volume 3364, Orlando, Florida, 1998.

Niel, Albert / Burgstaller, Markus: Der Roboter als Messmaschine zur 3D-Vermessung von komplex geformten Stahlteilen. 7. ABW-Workshop, Esslingen, 2002.

Niessner, Anton: Qualitative Deformationsanalyse unter Ausnützung der Farbinformation. PhD Thesis, Vienna University of Technology, 2002.

Paar, Gerhard / Bauer, Arnold: Cavity Surface Measuring System Using Stereo Reconstruction. In: Proceedings of SPIE Conference on Intelligent Robots and Computer Vision XV, Boston, 1996.

Paar, Gerhard / Rottensteiner, Franz / Pötzleitner, Wolfgang: Image Matching Strategies. In: Digital Image Analysis, Kropatsch / Bischof (Editors), 1st Edition, Springer, Berlin/Heidelberg/New York, 2001.

Petrou, Maria / Bosdogianni, Panagiota: Image Processing - The Fundamentals. 1st Edition, John Wiley and Sons, New York/Chichester/Brisbane/Toronto, 1999.

Pratt, William K.: Digital Image Processing. 1st Edition, John Wiley and Sons, New York/Chichester/Brisbane/Toronto, 1978.

Ramon, Jan / Bruynooghe, Maurice: A Polynomial Time Computable Metric between Point Sets. Acta Informatica, Volume 37/10, pp. 765-780, 2001.

Reiterer, Alexander: Entwicklung und Erprobung eines schrittmotorgesteuerten Digitalnivelliers. Diploma Thesis, Vienna University of Technology, 2001.

Reiterer, Alexander / Kahmen, Heribert / Egly, Uwe / Eiter, Thomas: 3D-Vermessung mit Videotheodoliten und automatisierte Zielerfassung mit Hilfe von Interest Operatoren. Allgemeine Vermessungsnachrichten (AVN), Volume 04, pp. 150-156, Herbert Wichmann, 2003.

Reiterer, Alexander / Kahmen, Heribert / Egly, Uwe / Eiter, Thomas: Knowledge-based Image Preprocessing for a Theodolit Measurement System. In: Optical 3-D Measurement Techniques VI, Grün / Kahmen (Editors), Volume I, pp. 183-190, ETH Zurich, 2003.

Reiterer, Alexander / Kahmen, Heribert / Egly, Uwe / Eiter, Thomas: Wissensbasierte Bildaufbereitung für ein Videotheodolit-basiertes Multisensorsystem. Submitted for publication (Allgemeine Vermessungsnachrichten - AVN), 2004.

Reiterer, Alexander / Kahmen, Heribert / Egly, Uwe / Eiter, Thomas: Knowledge-Based Videometric Measurement System. Submitted for publication (IFIP WG12.6 - First IFIP Conference on Artificial Intelligence Applications and Innovations), 2004.

Riekert, Wolf-Fritz / Gunther, Oliver / Hess, Gunter: Architecture of a Knowledge-Based System for Remote Sensor Data Analysis. In: Proceedings of the third International Conference on Industrial and Engineering Applications of Artificial Intelligence and Expert Systems, Volume I, pp. 228-232, ACM Press, 1990.

Rogers, Erika / DuPont, Versonya / Murphy, R. Robin / Warsi, Nazir: Knowledge-Based Image Enhancement for Cooperative Tele-Assistance. In: Proceedings of Symposium on Human Interaction with Complex Systems, Kluwer, 1995.

Roic, Miodrag: Erfassung von nicht signalisierten 3D-Strukturen mit Videotheodoliten. PhD Thesis, Vienna University of Technology, 1996.

Russel, Stuart J. / Norvig, Peter: Artificial Intelligence – A Modern Approach. 1st Edition, Prentice-Hall Inc., 1995.

Schmid, Cordelia / Mohr, Roger / Bauckhage, Christian: Evaluation of Interest Point Detectors. In: International Journal of Computer Vision, Volume 37/2, pp. 151-172, 2000.

SNNS-Project-page: http://www-ra.informatik.uni-tuebingen.de/SNNS, 2004.

Strecker, Stefan: Künstliche Neuronale Netze - Aufbau und Funktionsweise. Arbeitspapier - Lehrstuhl für allg. Wirtschaftsinformatik, Universität Mainz, Volume 10, 1997.

SAMBA-Endbericht – Projekt: Wissensbasiertes System 'Messtechnik im Bauwesen'. University of Hannover, 1995.

De Seixas, Andrea: 3D-Objektrekonstruktion mittels Gitterlinien-Verfahren. PhD Thesis, Vienna University of Technology, 2001.

Stefik, Mark: Introduction to Knowledge Systems. 2nd Edition, Morgan Kaufmann, San Francisco, 1998.

Walser, Bernd / Braunecker, Bernhard: Automation of Surveying Systems through Integration of Image Analysis Methods. In: Optical 3-D Measurement Techniques VI, Grün / Kahmen (Editors), Volume I, pp. 191-198, ETH Zurich, 2003.

Wasmeier, Peter: The Potential of Object Recognition Using a Servo-tacheometer TCA2003. In: Optical 3-D Measurement Techniques VI, Grün / Kahmen (Editors), Volume II, pp. 48-54, ETH Zurich, 2003.

von Webern, Hauke: Optimization and Extension of Gridline-methods to Detect Object Displacements and Deformations. In: Optical 3-D Measurement Techniques VI, Grün / Kahmen (Editors), Volume II, pp. 104-111, ETH Zurich, 2003.

WxClips-Project-page: http://www.anthemion.co.uk/wxclips, 2004.

Zell, Andreas: Simulation Neuronaler Netze. 1st Edition, Addison-Wesley Publishing, 1994.

Part VI

Appendix

Appendix A

Haralick Features

H_1 - **Angular Second Moment:**
$$H_1 = \sum_i \sum_j p(i,j)^2 \tag{A.1}$$

H_2 - **Contrast:**
$$H_2 = \sum_{n=0}^{N_g-1} n^2 \left\{ \sum_{i=1}^{N_g} \sum_{j=1}^{N_g} p(i,j) \right\}; \quad n = |i-j| \tag{A.2}$$

H_3 - **Correlation:**
$$H_3 = \frac{\sum_i \sum_j (ij) p(i,j) - \mu_x \mu_y}{\sigma_x \sigma_y} \tag{A.3}$$

H_4 - **Sum of Squares:**
$$H_4 = \sum_i \sum_j (i-\mu)^2 p(i,j) \tag{A.4}$$

H_5 - **Inverse Difference Moment:**
$$H_5 = \sum_i \sum_j \frac{1}{1+(i-j)^2} p(i,j) \tag{A.5}$$

H_6 - **Sum Average:**
$$H_6 = \sum_{i=2}^{2N_g} i p_{x+y}(i) \tag{A.6}$$

H_7 - **Sum Variance:**
$$H_7 = \sum_{i=2}^{2N_g} (i-f_g)^2 p_{x+y}(i) \tag{A.7}$$

H_8 - **Sum Entropy:**
$$H_8 = -\sum_{i=2}^{2N_g} p_{x+y}(i) \log\{p_{x+y}(i)\} = f_g \tag{A.8}$$

H_9 - **Entropy:**

$$H_9 = \sum_i \sum_j p(i,j) \log\{p(i,j)\} \tag{A.9}$$

H_{10} - **Difference Variance:**

$$H_{10} = \sum_{i=0}^{N_g-1} i^2 p_{x-y}(i) \tag{A.10}$$

H_{11} - **Difference Entropy:**

$$H_{11} = -\sum_{i=0}^{N_g-1} p_{x-y}(i) \log\{p_{x-y}(i)\} \tag{A.11}$$

H_{12} - **Measure of Correlation 1:**

$$H_{12} = \frac{HXY - HXY1}{max\{HX, HY\}} \tag{A.12}$$

H_{13} -**Measure of Correlation 2:**

$$H_{13} = \{1 - exp[-2(HXY2 - HXY)]\}^2 \tag{A.13}$$

with:

$$HXY = -\sum_i \sum_j p(i,j) \log\{p(i,j)\} \tag{A.14}$$

$$HXY1 = -\sum_i \sum_j p(i,j) \log\{p_x(i) p_y(j)\} \tag{A.15}$$

$$HXY2 = -\sum_i \sum_j p_x(i) p_y(j) \log\{p_x(i) p_y(j)\} \tag{A.16}$$

Appendix B

Membership Functions

M_1 - **Mean:**

0.000	$\leq$	very low	$\leq$	70.000
70.000	$<$	low	$\leq$	110.000
110.000	$<$	middle	$\leq$	146.000
146.000	$<$	high	$\leq$	186.000
186.000	$<$	very high	$\leq$	255.000

M_2 - **Standard Deviation:**

0.000	$\leq$	very low	$\leq$	25.000
25.000	$<$	low	$\leq$	30.000
30.000	$<$	middle	$\leq$	60.000
60.000	$<$	high	$\leq$	125.000
125.000	$<$	very high	$\leq$	255.000

M_3 - **Skew:**

		very high negative	$\leq$	-0.800
-0.800	$<$	high negative	$\leq$	-0.600
-0.600	$<$	middle negative	$\leq$	-0.400
-0.400	$<$	low negative	$\leq$	-0.200
-0.200	$<$	very low negative	$\leq$	0.000
0.000	$<$	very low positive	$\leq$	0.200
0.200	$<$	low positive	$\leq$	0.400
0.400	$<$	middle positive	$\leq$	0.600
0.600	$<$	high positive	$\leq$	0.800
0.800	$<$	very high positive		

H_1 - **Angular Second Moment:**

0.0000	$\leq$	very low	$\leq$	0.0004
0.0004	$<$	low	$\leq$	0.0006
0.0006	$<$	middle	$\leq$	0.0025
0.0025	$<$	high	$\leq$	0.0050
0.0050	$<$	very high		

H_2 - **Contrast:**

0.000	$\leq$	very low	$\leq$	50.000
50.000	$<$	low	$\leq$	100.000
100.000	$<$	middle	$\leq$	200.000
200.000	$<$	high	$\leq$	300.000
300.000	$<$	very high		

H_5 - **Inverse Difference Moment:**

0.000	$\leq$	very low	$\leq$	0.150
0.150	$<$	low	$\leq$	0.300
0.300	$<$	middle	$\leq$	0.400
0.400	$<$	high	$\leq$	0.600
0.600	$<$	very high		

H_9 - **Entropy:**

0.000	$\leq$	very low	$\leq$	2.000
2.000	$<$	low	$\leq$	3.000
3.000	$<$	middle	$\leq$	3.500
3.500	$<$	high	$\leq$	4.000
4.000	$<$	very high		

Appendix C

Knowledge Bases

C.1 Knowledge Base for Image Preprocessing and Image Enhancement

```
;******************************************************************************
; functions for log-file (to document conclusions).
;******************************************************************************

;------------------------------------------------------------------------------
; "rulename" gibt den Regelnamen.
;------------------------------------------------------------------------------
(deffunction rulename (?rulename)
   (open  ?*why_how_file* out2 "a")
   (printout out2 "*******************************************")
   (format out2 "%n")
   (printout out2 "rule: ") (printout out2 ?rulename)
   (format out2 "%n")
   (close out2)
)

;------------------------------------------------------------------------------
; "which" puts out factname and factvalue of fired rule.
;------------------------------------------------------------------------------
(deffunction which (?fact ?factname ?factvalue)
   (open  ?*why_how_file* out2 "a")
   (printout out2 ?fact)(printout out2 ?factname)(printout out2 "=")
   (printout out2 ?factvalue)
   (format out2 "%n")
   (close out2)
)

;------------------------------------------------------------------------------
; "why" puts out an explanation.
;------------------------------------------------------------------------------
(deffunction why (?why)
   (open  ?*why_how_file* out2 "a")
   (printout out2 "explanation:")
   (format out2 "%n")
   (printout out2 ?why)
```

```
    (format out2 "%n")
    (format out2 "%n")
    (close out2)
)

;------------------------------------------------------------------------
; "doc_user-interactions" documents the user-interactions.
;------------------------------------------------------------------------
(deffunction doc_user-interactions (?q)
    (open ?*why_how_file* out2 "a")
    (if (= ?q 0) then (printout out2 "user-interaction: -> no")
                 else (printout out2 "user-interaction: -> yes"))
    (format out2 "%n")
    (format out2 "%n")
    (close out2)
)

;------------------------------------------------------------------------
; "why_how-header" puts out a header.
;------------------------------------------------------------------------
(defrule why_how-header
    (declare (salience 300))
    =>
    (open ?*why_how_file* out2 "a")
    (format out2 "**********************************************")
    (format out2 "%n")
    (format out2 "Image preprocessing and image enhancement")
    (format out2 "%n")
    (format out2 "**********************************************")
    (format out2 "%n") (format out2 "%n")
    (close out2)
)

;************************************************************************
; rules for detection of homogeneity.
;************************************************************************

;------------------------------------------------------------------------
; "lh_h" detect the horizontal homogeneity.
;------------------------------------------------------------------------
(defrule lh_h
    (declare (salience 200))
    ?b <- (1_Haralick_Moment (H1_0_f very_high | high))
    ?c <- (5_Haralick_Moment (H5_0_f very_high | high))
    (not (1_Haralick_Moment (H1_90_f very_high | high)))
    (not (5_Haralick_Moment (H5_90_f very_high | high)))
=>
    (assert (condition (lh h))) (rulename "lh_h")
    (which ?b " H1_0_f" (fact-slot-value ?b H1_0_f))
    (which ?c " H5_0_f" (fact-slot-value ?c H5_0_f))
    (why "horizontal homogeneity.")
)

;------------------------------------------------------------------------
; "lh_v" detect the vertical homogeneity.
;------------------------------------------------------------------------
(defrule lh_v
```

C.1: Knowledge Base for Image Preprocessing and Image Enhancement

```
    (declare (salience 200))
    ?b <- (1_Haralick_Moment (H1_90_f very_high | high))
    ?c <- (5_Haralick_Moment (H5_90_f very_high | high))
    (not (1_Haralick_Moment (H1_0_f very_high | high)))
    (not (5_Haralick_Moment (H5_0_f very_high | high)))
=>
    (assert (condition (lh v)))  (rulename "lh_v")
    (which ?b " H1_90_f" (fact-slot-value ?b H1_90_f))
    (which ?c " H5_90_f" (fact-slot-value ?c H5_90_f))
    (why "vertical homogeneity.")
)
;************************************************************************
; rules for the choice of suitable algorithms.
;************************************************************************

;----------------------------------------------------------------------
; "np" detects an image with poor quality.
;----------------------------------------------------------------------
(defrule np
    (declare (salience 200))
    ?b <- (Stat_Moments (M1_f very_low | very_high))
    ?c <- (Stat_Moments (M2_f very_low))
    ?d <- (Stat_Moments (M3_f very_high_negative | very_high_positive))
=>
    (assert (condition (np yes)))
    (rulename "np")
    (which ?b " M1_f" (fact-slot-value ?b M1_f))
    (which ?c " M2_f" (fact-slot-value ?c M2_f))
    (which ?d " M3_f" (fact-slot-value ?d M3_f))
    (why "poor quality.")
)

;----------------------------------------------------------------------
; "thresholding" as postprocessing for edge detection.
;----------------------------------------------------------------------
(defrule thresholding
    (declare (salience 200))
    ?b <- (condition (np yes))
=>
    (assert (condition (thresholding yes)))
    (rulename "thresholding")
    (which ?b " np" (fact-slot-value ?b np))
    (why "thresholding as postprocessing for edge detection.")
)

;----------------------------------------------------------------------
; "histo" correct the contrast of an image (histogram equalization).
;----------------------------------------------------------------------
(defrule histo
    (declare (salience 200))
    ?b <- (Stat_Moments (M1_f very_low | low | very_high | high))
    ?d <- (Stat_Moments (M3_f low_positive | very_low_positive |
                              low_negative | very_low_negative))
 =>
    (assert (condition (histo yes)))  (rulename "histo")
    (which ?b " M1_f" (fact-slot-value ?b M1_f))
    (which ?d " M3_f" (fact-slot-value ?d M3_f))
```

```
      (why "correct the contrast of an image.")
)

;-------------------------------------------------------------------------------
; "ib" correct the contrast of an image (image brightening).
;-------------------------------------------------------------------------------
(defrule ib
    (declare (salience 200))
    ?b <- (Stat_Moments (M1_f very_low | low))
    ?d <- (Stat_Moments (M3_f middle_negative | high_negative | very_high_negative |
                         middle_positive | high_positive | very_high_positive))
=>
    (assert (condition (ib yes))) (rulename "ib")
    (which ?b " M1_f" (fact-slot-value ?b M1_f))
    (which ?d " M3_f" (fact-slot-value ?d M3_f))
    (why "image brightening.")
)
;-------------------------------------------------------------------------------
; "id" correct the contrast of an image (image darkening).
;-------------------------------------------------------------------------------
(defrule id
    (declare (salience 200))
    ?b <- (Stat_Moments (M1_f very_high | high))
    ?d <- (Stat_Moments (M3_f middle_positive | high_positive | very_high_positive |
                         middle_negative | high_negative | very_high_negative))

 =>
    (assert (condition (id yes))) (rulename "abdunkel")
    (which ?b " M1_f" (fact-slot-value ?b M1_f))
    (which ?d " M3_f" (fact-slot-value ?d M3_f))
    (why "image darkening.")
)

;-------------------------------------------------------------------------------
; "edge" makes a verification if edge detection is necessary
;-------------------------------------------------------------------------------
(defrule edge
    (declare (salience 200))
    (not (condition (kantenex yes)))
    ?b <- (Stat_Moments (M1_f very_low))
    ?c <- (Stat_Moments (M2_f very_low))
    ?d <- (Stat_Moments (M3_f middle_negative | high_negative | very_high_negative |
    middle_positive | high_positive | very_high_positive))
=>
    (assert (condition (edge yes))) (rulename "edge")
    (which ?b " M1_f" (fact-slot-value ?b M1_f))
    (which ?c " M2_f" (fact-slot-value ?c M2_f))
    (which ?d " M3_f" (fact-slot-value ?d M3_f))
    (why "edge detection is necessary.")
)

;-------------------------------------------------------------------------------
; "sobel_h" decides if sobel edge detection is necessary (horiz.).
;-------------------------------------------------------------------------------
(defrule sobel_h
    (declare (salience 200))
    (not (condition (np yes)))
```

C.1: Knowledge Base for Image Preprocessing and Image Enhancement

```
  ?b <- (condition (edge yes))
  ?c <- (condition (lh h))
  ?d <- (9_Haralick_Moment (H9_AVG_f very_low | low))
  =>
  (assert (condition (sobel yes)))
  (bind ?*edge_param* "SobelH") (rulename "sobel_h")
  (which ?b " condition_edge" (fact-slot-value ?b edge))
  (which ?c " condition_lh" (fact-slot-value ?c lh))
  (which ?d " H9_AVG_f" (fact-slot-value ?d H9_AVG_f))
  (why "Sobel is necessary.")
  (doc_user-interactions ?*edge_position*)
)

;-----------------------------------------------------------------
; "sobel_v" decides if sobel edge detection is necessary (vertic.).
;-----------------------------------------------------------------
(defrule sobel_v
  (declare (salience 200))
  (not (condition (np yes)))
  ?b <- (condition (edge yes))
  ?c <- (condition (lh v))
  ?d <- (9_Haralick_Moment (H9_AVG_f very_low | low))
  =>
  (assert (condition (sobel yes)))
  (bind ?*edge_param* "SobelV") (rulename "sobel_v")
  (which ?b " condition_edge" (fact-slot-value ?b edge))
  (which ?c " condition_lh" (fact-slot-value ?c lh))
  (which ?d " H9_AVG_f" (fact-slot-value ?d H9_AVG_f))
  (why "Sobel is necessary.")
  (doc_user-interactions ?*edge_position*)
)

;-----------------------------------------------------------------
; "prewitt_h" decides if prewitt edge detection is necessary (horiz.).
;-----------------------------------------------------------------
(defrule prewitt_h
  (declare (salience 200))
  (not (condition (np yes)))
  ?b <- (condition (edge yes))
  ?c <- (condition (lh h))
  ?d <- (9_Haralick_Moment (H9_AVG_f middle))
  =>
  (assert (condition (prewitt yes)))
  (bind ?*edge_param* "PrewittH") (rulename "prewitt_h")
  (which ?b " condition_kantenex" (fact-slot-value ?b edge))
  (which ?c " condition_lh" (fact-slot-value ?c lh))
  (which ?d " H9_AVG_f" (fact-slot-value ?d H9_AVG_f))
  (why "Prewitt is necessary.") (doc_user-interactions ?*edge_position*)
)

;-----------------------------------------------------------------
; "prewitt_v" decides if prewitt edge detection is necessary (vertic.).
;-----------------------------------------------------------------
(defrule prewitt_v
  (declare (salience 200))
  (not (condition (np yes)))
  ?b <- (condition (edge yes))
```

```
  ?c <- (condition (lh v))
  ?d <- (9_Haralick_Moment (H9_AVG_f middle))
  =>
  (assert (condition (prewitt yes)))
  (bind ?*edge_param* "PrewittV") (rulename "prewitt_v")
  (which ?b " condition_kantenex" (fact-slot-value ?b edge))
  (which ?c " condition_lh" (fact-slot-value ?c lh))
  (which ?d " H9_AVG_f" (fact-slot-value ?d H9_AVG_f))
  (why "Prewitt is necessary.") (doc_user-interactions ?*edge_position*)
)

;--------------------------------------------------------------------------------
; "laplace" decides if laplace edge detection is necessary.
;--------------------------------------------------------------------------------
(defrule laplace
  (not (condition (laplace yes)))
  ?b <- (condition (kantenex yes))
  (or (?c <- (condition (lh kl)))
      (and (condition (prewitt no))
           (condition (sobel no))))
=>
  (assert (condition (laplace yes)))
  (bind ?*edge_param* "Laplacian3x3")
  (rulename "laplace")
  (which ?b " condition_kantenex" (fact-slot-value ?b edge))
  (which ?c " condition_lh" (fact-slot-value ?c lh))
  (why "Laplace is necessary.")
  (doc_user-interactions ?*edge_position*)
)
;--------------------------------------------------------------------------------
; "gauss" decides if gauss filtering is necessary.
;--------------------------------------------------------------------------------
(defrule gauss
  (declare (salience 200))
  (not (condition (np yes)))
  ?b <- (condition (kantenex yes))
=>
  (assert (condition (gauss yes)))
  (bind ?*gauss_param* "Gaussian3x3")
  (rulename "gauss")
  (which ?b " condition_edge" (fact-slot-value ?b edge))
  (why "Gauss filtering is necessary.")
)

;--------------------------------------------------------------------------------
; "median" decides if median filtering is necessary.
;--------------------------------------------------------------------------------
(defrule median
  (declare (salience 200))
  (or (not (condition (edge yes)))
      (condition (np yes)))
=>
  (assert (condition (median yes)))
  (rulename "median")
  (why "Median filtering is necessary.")
  (doc_user-interactions ?*median_position*)
)
```

C.1: Knowledge Base for Image Preprocessing and Image Enhancement

```
;****************************************************************************
; rules to define the order of the algorithms.
;****************************************************************************
;----------------------------------------------------------------------------
; "histo_pos" fix the poition of "histo".
;----------------------------------------------------------------------------
(defrule histo_pos
   (declare (salience 100))
   (condition (histo yes))
=>
   (bind ?*position* (+ ?*position* 1)) (bind ?*histo_position* ?*position*)
)
;----------------------------------------------------------------------------
; "ib_pos" fix the poition of "ib".
;----------------------------------------------------------------------------
(defrule ib_pos
   (declare (salience 100))
   (condition (ib yes))
=>
   (bind ?*position* (+ ?*position* 1)) (bind ?*ib_id_position* ?*position*)
)
;----------------------------------------------------------------------------
; "id_pos" fix the poition of "id".
;----------------------------------------------------------------------------
(defrule id_pos
   (declare (salience 100))
   (condition (id yes))
=>
   (bind ?*position* (+ ?*position* 1)) (bind ?*ib_id_position* ?*position*)
)
;----------------------------------------------------------------------------
; "gauss_pos" fix the poition of "gauss".
;----------------------------------------------------------------------------
(defrule gauss_pos
   (declare (salience 95))
   (condition (gauss yes))
=>
   (bind ?*position* (+ ?*position* 1)) (bind ?*gauss_position* ?*position*)
)
;----------------------------------------------------------------------------
; "thresholding_pos" fix the poition of "thresholding".
;----------------------------------------------------------------------------
(defrule thresholding_pos
   (declare (salience 90))
   (condition (thresholding yes))
=>
   (bind ?*position* (+ ?*position* 1)) (bind ?*thresholding_position* ?*position*)
)
;----------------------------------------------------------------------------
; "median_onp_pos" fix the poition of "median" if np is no.
;----------------------------------------------------------------------------
```

```
(defrule median_onp_pos
  (declare (salience 80))
  (condition (median yes))
  (not (condition (np yes)))
=>
  (bind ?*position* (+ ?*position* 1)) (bind ?*median_position* ?*position*)
)

;----------------------------------------------------------------------------
; "median_onp_pos" fix the poition of "median" if np is yes.
;----------------------------------------------------------------------------
(defrule median_np_pos
  (declare (salience 70))
  (condition (median yes))
  (condition (np yes))
=>
  (bind ?*position* (+ ?*position* 1)) (bind ?*median_position* ?*position*)
)

;----------------------------------------------------------------------------
; "edge_onp_pos" fix the poition of "edge" if np is no.
;----------------------------------------------------------------------------
(defrule kantenex_onp_pos
  (declare (salience 60))
  (condition (edge yes))
  (not (condition (np yes)))
=>
  (bind ?*position* (+ ?*position* 1)) (bind ?*edge_position* ?*position*)
)

;----------------------------------------------------------------------------
; "edge_np_pos" fix the poition of "edge" if np is yes.
;----------------------------------------------------------------------------
(defrule kantenex_np_pos
  (declare (salience 50))
  (condition (edge yes))
  (condition (np yes))
=>
  (bind ?*position* (+ ?*position* 1)) (bind ?*edge_position* ?*position*)
)

;----------------------------------------------------------------------------
; "save" fix the poition of "save".
;----------------------------------------------------------------------------
(defrule save
  (declare (salience 40))
  =>
  (bind ?*position* (+ ?*position* 1)) (bind ?*save_position* ?*position*)
)

;****************************************************************************
; rules for the prededefinition of necessary parameters.
;****************************************************************************

;----------------------------------------------------------------------------
; "median_param" fix the parameter for median filtering.
;----------------------------------------------------------------------------
```

C.1: Knowledge Base for Image Preprocessing and Image Enhancement 131

```
(defrule median_param
  (condition (median yes))
=>
  (bind ?*median_param* 3)
)

;---------------------------------------------------------------
; "gauss_param" fix the parameter for gauss filtering.
;---------------------------------------------------------------
(defrule gauss_param
  (condition (gauss yes))
=>
  (bind ?*gauss_param* 3)
)

;---------------------------------------------------------------
; ib_param fix the parameter for ib.
;---------------------------------------------------------------
(defrule ib_param
  (condition (aufhell yes))
=>
  (bind ?x (fact-slot-value 1 M1))
  (bind ?*ib_id_param* (/ 128 ?x))
)

;---------------------------------------------------------------
; id_param fix the parameter for id.
;---------------------------------------------------------------
(defrule id_param
  (condition (abdunkel yes))
  =>
  (bind ?x (fact-slot-value 1 M1))
  (bind ?*ib_id_param* (/ 128 ?x))
)

;---------------------------------------------------------------
; thresholding_param fix the parameter for thresholding.
;---------------------------------------------------------------
(defrule thresholding_param
  (condition (schwellwert yes))
=>
  (bind ?*thresholding_param* (fact-slot-value 1 M1))
)


;***************************************************************
; rules for user-interaction.
;***************************************************************

;---------------------------------------------------------------
; user-interaction for "edge".
;---------------------------------------------------------------
(defrule edge_ui
  (declare (salience 5))
  (condition (edge yes))
=>
  (if (eq (lowcase(get-text-from-user "The kbs has detected that
```

C.1: Knowledge Base for Image Preprocessing and Image Enhancement

```
                    edge detection is necessary.)
                   Do edge detection (yes/no)?)) "no") then (bind ?*edge_position* 0))
)

;----------------------------------------------------------------------
; user-interaction for "median".
;----------------------------------------------------------------------
(defrule median_ui
   (declare (salience 5))
   (condition (median yes))
=>
   (if (eq (lowcase(get-text-from-user "The kbs has detected that
              median filtering is necessary.)
                    Do median filtering (yes/no)?")) "no") then (bind ?*median_position* 0))
)

;**********************************************************************
; Output
;**********************************************************************
(defrule output
    (declare (salience 0))
=>
   (open ?*input* input "r")  ; input from "image.ini"
   (bind ?imagefile (read input))
   (open ?*outputfile* out "a")

   (if (= ?*histo_position* 1) then
       (format out "[filter0]") (format out "%n") (format out "FilterType=")
       (format out "HistoEqualize") (format out "%n") (format out "param=")
       (format out "%n"))

   (if (= ?*ib_id_position* 1) then
       (format out "[filter0]") (format out "%n") (format out "FilterType=")
       (format out "MultiplySScale") (format out "%n") (format out "param=")
       (format out "%g" ?*ib_id_param*) (format out "%n"))

   (if (= ?*gauss_position* 1) then (format out "[filter0]") (format out "%n")
       (format out "FilterType=") (format out "FixedFilter") (format out "%n")
       (format out "param=") (format out ?*gauss_param*) (format out "%n"))

   (if (= ?*thresholding_position* 1) then (format out "[filter0]") (format out "%n")
       (format out "FilterType=") (format out "Threshold") (format out "%n")
       (format out "param=") (format out "%g" ?*thresholding_param*) (format out "%n"))

   (if (= ?*median_position* 1) then (format out "[filter0]") (format out "%n")
       (format out "FilterType=") (format out "MedianFilter") (format out "%n")
       (format out "param=" ) (format out "%g" ?*median_param*) (format out "%n"))

   (if (= ?*edge_position* 1) then (format out "[filter0]") (format out "%n")
       (format out "FilterType=") (format out "FixedFilter") (format out "%n")
       (format out "param=") (format out ?*edge_param*) (format out "%n"))

   (if (= ?*save_position* 1) then (format out "[filter0]") (format out "%n")
       (format out "FilterType=") (format out "save image") (format out "%n")
       (format out "param=") (format out "%s" ?imagefile) (format out "_BVV.bmp")
```

```
            (format out "%n"))

....and so on.

  (close input)
  (close out)
)
```

C.2   Knowledge Base for Point Detection

```
;*****************************************************************************
; rules for the choice of suitable IOP.
;*****************************************************************************

;-----------------------------------------------------------------------------
; "HFVM" decides if the HFVM operator is the best one.
;-----------------------------------------------------------------------------
(defrule HFVM
  (declare (salience 200))
  (not (iop (HFVM yes)))
  (and (and (UQ (FT D))
            (UQ (RFL no)))
       (UQ (S no)))
=>
  (assert (iop (HFVM yes)))
  (rulename "HFVM")
  (why "HFVM-Operator is suited for steel glass facades.")
)

;-----------------------------------------------------------------------------
; "HRS" decides if the Harris operator is the best one.
;-----------------------------------------------------------------------------
(defrule HRS
  (declare (salience 200))
  (not (iop (HRS yes)))
  (or (and (1_Haralick_Moment (H1_0_f very_high | high | middle))
           (5_Haralick_Moment (H5_0_f very_high | high | middle)))
      (and (1_Haralick_Moment (H1_90_f very_high | high | middle))
           (5_Haralick_Moment (H5_90_f very_high | high | middle))))
=>
  (assert (iop (HRS yes)))
  (rulename "HRS")
  (why "HRS-Operator is suited for middle/high horizontal homogeneity.")
)

;-----------------------------------------------------------------------------
; "F2_SG" decides if the F2 operator is the best one (for steel glass facades).
;-----------------------------------------------------------------------------
(defrule F2_SG
  (declare (salience 200))
  (not (iop (HFVM yes)))
  (UQ (FT D))
```

```
  =>
  (assert (iop (F2 yes)))
  (rulename "F2_SG")
  (why "F2-Operator for steel glass facades (if HFVM is not suited."))
)

;--------------------------------------------------------------------------------
; F2 decides if the F2 operator is the best one.
;--------------------------------------------------------------------------------
(defrule F2
  (declare (salience 200))
  (not (iop (F2 yes)))
  (not (iop (F1 yes)))
  (or (or (or (and (1_Haralick_Moment (H1_0_f low | very_low | middle))
                   (5_Haralick_Moment (H5_0_f low | very_low | middle)))
              (and (1_Haralick_Moment (H1_90_f low | very_low | middle))
                   (5_Haralick_Moment (H5_90_f low | very_low | middle))))
          (and (and (1_Haralick_Moment (H1_0_f low | very_low))
                    (1_Haralick_Moment (H1_90_f low | very_low)))
               (not (Stat_Moments (M3_f very_high_negativ)))))
      (and (and (5_Haralick_Moment (H5_0_f low | very_low))
                (5_Haralick_Moment (H5_90_f low | very_low)))
           (not (Stat_Moments (M3_f very_high_negativ)))))
=>
  (assert (iop (F2 yes)))
  (rulename "F2")
  (why "F2 operator is suited for no/middle horizontal homogeneity."))
)

;--------------------------------------------------------------------------------
; F1 decides if the F1 operator is the best one.
;--------------------------------------------------------------------------------
(defrule F1
  (declare (salience 200))
  (not (iop (F1 yes)))
  (not (iop (F2 yes)))
  (or (or (or (and (1_Haralick_Moment (H1_0_f low | very_low | middle))
                   (5_Haralick_Moment (H5_0_f low | very_low | middle)))
              (and (1_Haralick_Moment (H1_90_f low | very_low | middle))
                   (5_Haralick_Moment (H5_90_f low | very_low | middle))))
          (and (and (1_Haralick_Moment (H1_0_f low | very_low))
                    (1_Haralick_Moment (H1_90_f low | very_low)))
               (Stat_Moments (M3_f very_high_negativ))))
      (and (and (5_Haralick_Moment (H5_0_f low | very_low))
                (5_Haralick_Moment (H5_90_f low | very_low)))
           (Stat_Moments (M3_f very_high_negativ))))
=>
  (assert (iop (F1 yes)))
  (rulename "F1")
  (why "F1 operator is suited for no/middle horizontal homogeneity."))
)

;********************************************************************************
; rules to define the order of the IOP.
;********************************************************************************
```

C.2: Knowledge Base for Point Detection 135

```
;--------------------------------------------------------------------------------
; "HFVM_pos" fix the poition of "HFVM".
;--------------------------------------------------------------------------------
(defrule HFVM_pos
  (declare (salience 100))
  (iop (HFVM yes))
=>
  (bind ?*position_iop* (+ ?*position_iop* 1))
  (bind ?*HFVM_position* ?*position_iop*)
)

;--------------------------------------------------------------------------------
; "F1_pos" fix the poition of "F1".
;--------------------------------------------------------------------------------
(defrule F1_pos
  (declare (salience 90))
  (iop (F1 yes))
=>
  (bind ?*position_iop* (+ ?*position_iop* 1))
  (bind ?*F1_position* ?*position_iop*)
)

;--------------------------------------------------------------------------------
; "F2_pos" fix the poition of "F2".
;--------------------------------------------------------------------------------
(defrule F2_pos
  (declare (salience 90))
  (iop (F2 yes))
=>
  (bind ?*position_iop* (+ ?*position_iop* 1))
  (bind ?*F2_position* ?*position_iop*)
)

;--------------------------------------------------------------------------------
; "HRS_pos" fix the poition of "HRS".
;--------------------------------------------------------------------------------
(defrule HRS_pos
  (declare (salience 80))
  (iop (HRS yes))
=>
  (bind ?*position_iop* (+ ?*position_iop* 1))
  (bind ?*HRS_position* ?*position_iop*)
)

;********************************************************************************
; rules for the prededefinition of necessary parameters for IOP.
;********************************************************************************

;--------------------------------------------------------------------------------
; "HFVM_param" fix the parameter for "HFVM".
;--------------------------------------------------------------------------------
(defrule HFVM_param
  (declare (salience 10))
  (iop (HFVM yes))
=>
  (bind ?*HFVM_param* "Feat.cfg")
)
```

```
;---------------------------------------------------------------------------
; "F1_param" fix the parameter for "F1".
;---------------------------------------------------------------------------
(defrule F1_param
  (declare (salience 10))
  (iop (F1 yes))
=>
  (bind ?*F1_param* 3)
)

;---------------------------------------------------------------------------
; "HRS_param" fix the parameter for "HRS".
;---------------------------------------------------------------------------
(defrule HRS_param
  (declare (salience 10))
  (iop (HRS yes))
=>
  (bind ?M1 (fact-slot-value 1 M1_f))
  (bind ?H2 (fact-slot-value 3 H2_AVG_f))
  (if (and (= (str-compare ?M1 "very_low") 0) (= (str-compare ?H2 "very_low") 0))
    then (bind ?*HRS_param* 0.00005))
  (if (and (= (str-compare ?M1 "very_low") 0) (= (str-compare ?H2 "low") 0))
    then (bind ?*HRS_param* 0.0001))
  (if (and (= (str-compare ?M1 "very_low") 0) (= (str-compare ?H2 "middle") 0))
    then (bind ?*HRS_param* 0.00015))
  (if (and (= (str-compare ?M1 "very_low") 0) (= (str-compare ?H2 "high") 0))
    then (bind ?*HRS_param* 0.00023))
  (if (and (= (str-compare ?M1 "very_low") 0) (= (str-compare ?H2 "low") 0))
    then (bind ?*HRS_param* 0.0003))
  (if (and (= (str-compare ?M1 "low") 0) (= (str-compare ?H2 "very_low") 0))
    then (bind ?*HRS_param* 0.00006))
  (if (and (= (str-compare ?M1 "low") 0) (= (str-compare ?H2 "low") 0))
    then (bind ?*HRS_param* 0.00095))
  (if (and (= (str-compare ?M1 "low") 0) (= (str-compare ?H2 "middle") 0))
    then (bind ?*HRS_param* 0.0018))
  (if (and (= (str-compare ?M1 "low") 0) (= (str-compare ?H2 "high") 0))
    then (bind ?*HRS_param* 0.0028))
  (if (and (= (str-compare ?M1 "low") 0) (= (str-compare ?H2 "low") 0))
    then (bind ?*HRS_param* 0.0037))
  (if (and (= (str-compare ?M1 "middle") 0) (= (str-compare ?H2 "very_low") 0))
    then (bind ?*HRS_param* 0.00007))
  (if (and (= (str-compare ?M1 "middle") 0) (= (str-compare ?H2 "low") 0))
    then (bind ?*HRS_param* 0.0018))
  (if (and (= (str-compare ?M1 "middle") 0) (= (str-compare ?H2 "middle") 0))
    then (bind ?*HRS_param* 0.0035))
  (if (and (= (str-compare ?M1 "middle") 0) (= (str-compare ?H2 "high") 0))
    then (bind ?*HRS_param* 0.0053))
  (if (and (= (str-compare ?M1 "middle") 0) (= (str-compare ?H2 "low") 0))
    then (bind ?*HRS_param* 0.007))
  (if (and (= (str-compare ?M1 "high") 0) (= (str-compare ?H2 "very_low") 0))
    then (bind ?*HRS_param* 0.000085))
  (if (and (= (str-compare ?M1 "high") 0) (= (str-compare ?H2 "low") 0))
    then (bind ?*HRS_param* 0.0022))
  (if (and (= (str-compare ?M1 "high") 0) (= (str-compare ?H2 "middle") 0))
    then (bind ?*HRS_param* 0.00425))
  (if (and (= (str-compare ?M1 "high") 0) (= (str-compare ?H2 "high") 0))
```

C.2: Knowledge Base for Point Detection 137

```
      then (bind ?*HRS_param* 0.0064))
    (if (and (= (str-compare ?M1 "high") 0) (= (str-compare ?H2 "low") 0))
      then (bind ?*HRS_param* 0.0085))
    (if (and (= (str-compare ?M1 "very_high") 0) (= (str-compare ?H2 "very_low") 0))
      then (bind ?*HRS_param* 0.0001))
    (if (and (= (str-compare ?M1 "very_high") 0) (= (str-compare ?H2 "low") 0))
      then (bind ?*HRS_param* 0.0026))
    (if (and (= (str-compare ?M1 "very_high") 0) (= (str-compare ?H2 "middle") 0))
      then (bind ?*HRS_param* 0.005))
    (if (and (= (str-compare ?M1 "very_high") 0) (= (str-compare ?H2 "high") 0))
      then (bind ?*HRS_param* 0.0075))
    (if (and (= (str-compare ?M1 "very_high") 0) (= (str-compare ?H2 "low") 0))
      then (bind ?*HRS_param* 0.01))
)

;------------------------------------------------------------------------------
; "F2_param" fix the parameter for "F2".
;------------------------------------------------------------------------------
(defrule F2_param
    (declare (salience 10))
    (iop (F2 yes))
=>
    (bind ?M1 (fact-slot-value 1 M1_f))
    (bind ?H2 (fact-slot-value 3 H2_AVG_f))
    (if (and (= (str-compare ?M1 "very_low") 0) (= (str-compare ?H2 "very_low") 0))
      then (bind ?*F2_param* 32))
    (if (and (= (str-compare ?M1 "very_low") 0) (= (str-compare ?H2 "low") 0))
      then (bind ?*F2_param* 36))
    (if (and (= (str-compare ?M1 "very_low") 0) (= (str-compare ?H2 "middle") 0))
      then (bind ?*F2_param* 40))
    (if (and (= (str-compare ?M1 "very_low") 0) (= (str-compare ?H2 "high") 0))
      then (bind ?*F2_param* 45))
    (if (and (= (str-compare ?M1 "very_low") 0) (= (str-compare ?H2 "low") 0))
      then (bind ?*F2_param* 50))
    (if (and (= (str-compare ?M1 "low") 0) (= (str-compare ?H2 "very_low") 0))
      then (bind ?*F2_param* 176))
    (if (and (= (str-compare ?M1 "low") 0) (= (str-compare ?H2 "low") 0))
      then (bind ?*F2_param* 198))
    (if (and (= (str-compare ?M1 "low") 0) (= (str-compare ?H2 "middle") 0))
      then (bind ?*F2_param* 220))
    (if (and (= (str-compare ?M1 "low") 0) (= (str-compare ?H2 "high") 0))
      then (bind ?*F2_param* 248))
    (if (and (= (str-compare ?M1 "low") 0) (= (str-compare ?H2 "low") 0))
      then (bind ?*F2_param* 175))
    (if (and (= (str-compare ?M1 "middle") 0) (= (str-compare ?H2 "very_low") 0))
      then (bind ?*F2_param* 320))
    (if (and (= (str-compare ?M1 "middle") 0) (= (str-compare ?H2 "low") 0))
      then (bind ?*F2_param* 360))
    (if (and (= (str-compare ?M1 "middle") 0) (= (str-compare ?H2 "middle") 0))
      then (bind ?*F2_param* 400))
    (if (and (= (str-compare ?M1 "middle") 0) (= (str-compare ?H2 "high") 0))
      then (bind ?*F2_param* 450))
    (if (and (= (str-compare ?M1 "middle") 0) (= (str-compare ?H2 "low") 0))
      then (bind ?*F2_param* 500))
    (if (and (= (str-compare ?M1 "high") 0) (= (str-compare ?H2 "very_low") 0))
      then (bind ?*F2_param* 480))
    (if (and (= (str-compare ?M1 "high") 0) (= (str-compare ?H2 "low") 0))
```

```
      then (bind ?*F2_param* 540))
(if (and (= (str-compare ?M1 "high") 0) (= (str-compare ?H2 "middle") 0))
      then (bind ?*F2_param* 600))
(if (and (= (str-compare ?M1 "high") 0) (= (str-compare ?H2 "high") 0))
      then (bind ?*F2_param* 675))
(if (and (= (str-compare ?M1 "high") 0) (= (str-compare ?H2 "low") 0))
      then (bind ?*F2_param* 750))
(if (and (= (str-compare ?M1 "very_high") 0) (= (str-compare ?H2 "very_low") 0))
      then (bind ?*F2_param* 640))
(if (and (= (str-compare ?M1 "very_high") 0) (= (str-compare ?H2 "low") 0))
      then (bind ?*F2_param* 720))
(if (and (= (str-compare ?M1 "very_high") 0) (= (str-compare ?H2 "middle") 0))
      then (bind ?*F2_param* 800))
(if (and (= (str-compare ?M1 "very_high") 0) (= (str-compare ?H2 "high") 0))
      then (bind ?*F2_param* 900))
(if (and (= (str-compare ?M1 "very_high") 0) (= (str-compare ?H2 "low") 0))
      then (bind ?*F2_param* 1000))
)


;******************************************************************************
; Output
;******************************************************************************
(defrule ausgabe
  (declare (salience 0))
=>
  (open ?*outputfile2* out "a")
  (if (= ?*HFVM_position* 1) then (format out "interestpoints ")
      (format out "%s" ?imagefile) (format out " -Featurevector ")
      (format out "%s " ?*HFVM_param*) (format out " -Graphics ")
      (format out "%s" ?imagefile) (format out "_HFVM") (format out " -Maxpoints ")
      (format out "%g" ?*max_points*) (format out " -Mindist ")
      (format out "%g" ?*min_dist*) (format out " -SubListfile ")
      (format out "%s" ?imagefile) (format out "_SUB_HFVM.txt") (format out "%n")
      (format outiop "%s" ?imagefile) (format outiop "_SUB_HFVM.txt")
      (format outiop "%n") (format out "interestpoints ")
      (format out "%s" ?imagefile2)
      (format out " -Featurevector ") (format out "%s " ?*HFVM_param*)
      (format out " -Graphics ")   (format out "%s" ?imagefile2) (format out "_HFVM")
      (format out " -Maxpoints ") (format out "%g" ?*max_points*)
      (format out " -Mindist ")    (format out "%g" ?*min_dist*)
      (format out " -SubListfile ") (format out "%s" ?imagefile2)
      (format out "_SUB_HFVM.txt") (format out "%n"))

  (if (= ?*HRS_position* 1) then (format out "interestpoints ")
      (format out "%s" ?imagefile) (format out " -Harris ")
      (format out "%g " ?*HRS_scale*) (format out "%g " ?*HRS_sigma*)
      (format out "%g" ?*HRS_param*) (format out " -Graphics ")
      (format out "%s" ?imagefile) (format out "_HRS") (format out " -Maxpoints ")
      (format out "%g" ?*max_points*) (format out " -Mindist ")
      (format out "%g" ?*min_dist*) (format out " -SubListfile ")
      (format out "%s" ?imagefile) (format out "_SUB_HRS.txt")
      (format out "%n") (format outiop "%s" ?imagefile) (format outiop "_SUB_HRS.txt")
      (format outiop "%n") (format out "interestpoints ")
      (format out "%s" ?imagefile2) (format out " -Harris ")
      (format out "%g " ?*HRS_scale*) (format out "%g " ?*HRS_sigma*)
      (format out "%g" ?*HRS_param*) (format out " -Graphics ")
```

```
        (format out "%s" ?imagefile2) (format out "_HRS") (format out " -Maxpoints ")
        (format out "%g" ?*max_points*) (format out " -Mindist ")
        (format out "%g" ?*min_dist*) (format out " -SubListfile ")
        (format out "%s" ?imagefile2) (format out "_SUB_HRS.txt") (format out "%n"))

....and so on.

  (close out)
)
```

C.3   Knowledge Base for Point Filtering

```
;--------------------------------------------------------------------------------
; "HRS_w" weighted points detected by the Harris operator
;--------------------------------------------------------------------------------
(defrule HRS_w
  (declare (salience 9))
  ?f<-(Matching_1 (lfndNr_0 ?lfNr)
                  (IOP ?iop&: (= (str-compare ?iop "HRS") 0))
                  (param1 ?p1&: (> ?p1 0.1)) (param2 ?p2) (param3 ?p3) (param4 ?p4)
                  (X_0 ?x_0) (Y_0 ?y_0)
                  (X_M ?x_m) (Y_M ?y_m)
                  (lfndNr_1 ?lfNr1)
                  (X_1 ?x_1&: (<> ?x_1 -1)) (Y_1 ?y_1)
                  (gew ?gew)
                  (status_match ?sm) (status_hrs ?hrs&: (<> ?hrs 1)))
=>
  (retract ?f)
  (bind ?g (+ ?gew 2))

  (assert (Matching_1 (lfndNr_0 ?lfNr)
                      (IOP ?iop)
                      (param1 ?p1) (param2 ?p2) (param3 ?p3) (param4 ?p4)
                      (X_0 ?x_0) (Y_0 ?y_0)
                      (X_M ?x_m) (Y_M ?y_m)
                      (lfndNr_1 ?lfNr1)
                      (X_1 ?x_1) (Y_1 ?y_1)
                      (gew ?g) (status_match ?sm) (status_hrs 1)))
)

;--------------------------------------------------------------------------------
; "HFVM_w" weighted points detected by the HFVM operator
;--------------------------------------------------------------------------------
(defrule HFVM_w
  (declare (salience 8))
  ?f<-(Matching_1 (lfndNr_0 ?lfNr)
                  (IOP ?iop&: (= (str-compare ?iop "HFVM") 0))
                  (param1 ?p1&: (> ?p1 0.1)) (param2 ?p2) (param3 ?p3) (param4 ?p4)
                  (X_0 ?x_0) (Y_0 ?y_0)
                  (X_M ?x_m) (Y_M ?y_m)
                  (lfndNr_1 ?lfNr1)
                  (X_1 ?x_1&: (<> ?x_1 -1)) (Y_1 ?y_1)
```

```
                              (gew ?gew)
(status_match ?sm) (status_hrs ?hrs) (status_hfvm ?hfvm&: (<> ?hfvm 1)))
=>
   (retract ?f) ;entfernt den fact f
   (bind ?g (+ ?gew 2))
   (assert (Matching_1 (lfndNr_0 ?lfNr)
                       (IOP ?iop)
                       (param1 ?p1) (param2 ?p2) (param3 ?p3) (param4 ?p4)
                       (X_0 ?x_0) (Y_0 ?y_0)
                       (X_M ?x_m) (Y_M ?y_m)
                       (lfndNr_1 ?lfNr1)
                       (X_1 ?x_1) (Y_1 ?y_1)
                       (gew ?g) (status_match ?sm) (status_hrs ?hrs) (status_hfvm 1)))
)

;---------------------------------------------------------------------------------
; "F1_w" weighted points detected by the F1 operator
;---------------------------------------------------------------------------------
(defrule F1_w
   (declare (salience 7))
   ?f<-(Matching_1 (lfndNr_0 ?lfNr)
                   (IOP ?iop&: (= (str-compare ?iop "F1") 0))
                   (param1 ?p1&: (> ?p1 0.01)) (param2 ?p2) (param3 ?p3) (param4 ?p4)
                   (X_0 ?x_0) (Y_0 ?y_0)
                   (X_M ?x_m) (Y_M ?y_m)
                   (lfndNr_1 ?lfNr1)
                   (X_1 ?x_1&: (<> ?x_1 -1)) (Y_1 ?y_1)
                   (gew ?gew)
(status_match ?sm) (status_hrs ?hrs) (status_hfvm ?hfvm) (status_f1 ?f1&: (<> ?f1 1)))
=>
   (retract ?f)
   (bind ?g (+ ?gew 2))
   (assert (Matching_1 (lfndNr_0 ?lfNr)
                       (IOP ?iop)
                       (param1 ?p1) (param2 ?p2) (param3 ?p3) (param4 ?p4)
                       (X_0 ?x_0) (Y_0 ?y_0)
                       (X_M ?x_m) (Y_M ?y_m)
                       (lfndNr_1 ?lfNr1)
                       (X_1 ?x_1) (Y_1 ?y_1)
                       (gew ?g) (status_match ?sm) (status_hrs ?hrs) (status_hfvm ?hfvm)
                       (status_f1 1)))
)

;---------------------------------------------------------------------------------
; "F2_w" weighted points detected by the F2 operator
;---------------------------------------------------------------------------------
(defrule F2_w
   (declare (salience 6))
   ?f<-(Matching_1 (lfndNr_0 ?lfNr)
                   (IOP ?iop&: (= (str-compare ?iop "F2") 0))
(param1 ?p1) (param2 ?p2&: (> ?p2 1000)) (param3 ?p3) (param4 ?p4)
(X_0 ?x_0) (Y_0 ?y_0)
(X_M ?x_m) (Y_M ?y_m)
(lfndNr_1 ?lfNr1)
(X_1 ?x_1&: (<> ?x_1 -1)) (Y_1 ?y_1)
(gew ?gew)
(status_match ?sm)
```

C.3: Knowledge Base for Point Filtering

```
                        (status_hrs ?hrs)
                        (status_hfvm ?hfvm)
                        (status_f1 ?f1)
                        (status_f2 ?f2&: (<> ?f2 1)))
=>
  (retract ?f)
  (bind ?g (+ ?gew 2))
  (assert (Matching_1 (lfndNr_0 ?lfNr)
                      (IOP ?iop)
                      (param1 ?p1) (param2 ?p2) (param3 ?p3) (param4 ?p4)
                      (X_0 ?x_0) (Y_0 ?y_0)
                      (X_M ?x_m) (Y_M ?y_m)
                      (lfndNr_1 ?lfNr1)
                      (X_1 ?x_1) (Y_1 ?y_1)
                      (gew ?g)
                      (status_match ?sm)
                      (status_hrs ?hrs)
                      (status_hfvm ?hfvm)
                      (status_f1 ?f1)
                      (status_f2 1)))
)


;****************************************************************************
; Output filtered point list
;****************************************************************************

(defrule output
   (declare (salience 0))
   (Matching_1 (lfndNr_0 ?lfNr)
               (IOP ?iop)
               (param1 ?p1) (param2 ?p2) (param3 ?p3) (param4 ?p4)
               (X_0 ?x_0) (Y_0 ?y_0)
               (X_M ?x_m) (Y_M ?y_m)
               (lfndNr_1 ?lfNr1)
               (X_1 ?x_1) (Y_1 ?y_1)
               (gew ?gew))
=>
  (open ?*FromMatching* outfm "a")
  (format outfm "%g " ?lfNr) (format outfm "%s " ?iop)
  (format outfm "%g " ?p1) (format outfm "%g " ?p2)


....and so on.

  (close outfm)
)
```

Südwestdeutscher Verlag
für Hochschulschriften

Wissenschaftlicher Buchverlag bietet
kostenfreie
Publikation
von
Dissertationen und Habilitationen

Sie verfügen über eine wissenschaftliche Abschlußarbeit zu aktuellen oder zeitlosen Fragestellungen, die hohen inhaltlichen und formalen Anspruchen genügt, und haben **Interesse an einer honorarvergüteten Publikation?**

Dann senden Sie bitte erste Informationen über Ihre Arbeit per Email an:
info@svh-verlag.de.

Unser Außenlektorat meldet sich umgehend bei Ihnen.

Südwestdeutscher Verlag für Hochschulschriften
Aktiengesellschaft & Co. KG
Dudweiler Landstr. 99
D – 66123 Saarbrücken
www.svh-verlag.de

Printed by Books on Demand GmbH, Norderstedt / Germany